Magnus Rueping
Dixit Parmar
Erli Sugiono

Asymmetric Brønsted Acid Catalysis

Related Titles

Dalko, P.I. (ed.)

Comprehensive Enantioselective Organocatalysis

Catalysts, Reactions, and Applications

2013
Print ISBN: 978-3-527-33236-6; also available in electronic formats

Rios Torres, R. (ed.)

Stereoselective Organocatalysis

Bond Formation Methodologies and Activation Modes

2013
Print ISBN: 978-1-118-20353-8; also available in electronic formats

Koskinen, A.A.

Asymmetric Synthesis of Natural Products 2e

2 Edition

2012
Print ISBN: 978-1-119-97668-4; also available in electronic formats

Christmann, M., Bräse, S. (eds.)

Asymmetric Synthesis II

More Methods and Applications

2012
Print ISBN: 978-3-527-32921-2; also available in electronic formats

Cossy, J., Arseniyadis, S. (eds.)

Modern Tools for the Synthesis of Complex Bioactive Molecules

2012
Print ISBN: 978-0-470-61618-5; also available in electronic formats

Gruttadauria, M., Giacalone, F. (eds.)

Catalytic Methods in Asymmetric Synthesis

Advanced Materials, Techniques, and Applications

2011
Print ISBN: 978-0-470-64136-1; also available in electronic formats

Magnus Rueping, Dixit Parmar, and Erli Sugiono

Asymmetric Brønsted Acid Catalysis

WILEY-VCH
Verlag GmbH & Co. KGaA

Authors

Prof. Dr. Magnus Rueping
RWTH Aachen University
Institute of Organic Chemistry
Landoltweg 1
52074 Aachen
Germany

Dr. Dixit Parmar
RWTH Aachen University
Institute of Organic Chemistry
Landoltweg 1
52074 Aachen
Germany

Dr. Erli Sugiono
RWTH Aachen University
Institute of Organic Chemistry
Landoltweg 1
52074 Aachen
Germany

Library of Congress Card No.: applied for

British Library Cataloguing-in-Publication Data
A catalogue record for this book is available from the British Library.

Bibliographic information published by the Deutsche Nationalbibliothek
The Deutsche Nationalbibliothek lists this publication in the Deutsche Nationalbibliografie; detailed bibliographic data are available on the Internet at <http://dnb.d-nb.de>.

Print ISBN: 978-3-527-33917-4
ePDF ISBN: 978-3-527-69477-8
ePub ISBN: 978-3-527-69479-2
Mobi ISBN: 978-3-527-69480-8
oBook ISBN: 978-3-527-69478-5

Cover Design Formgeber, Mannheim, Germany
Typesetting SPi Global, Chennai, India
Printing and Binding Markono Print Media Pte Ltd

Printed on acid-free paper

Contents

Preface

Asymmetric Brønsted acid catalysis can be considered as one of the fundamental pillars of organocatalysis. Over the course of the last 10 years, it has become one of the most hotly researched fields in modern organic chemistry. Unlike certain fads in chemistry, which seem to gather fame as fast as they lose it, chiral Brønsted acids appear to be well equipped to withstand fashion trends and now hold a permanent position inside synthetic organic toolboxes. Since appearing on the scene in 2004, the number of published reports in this area has been steadily increasing and the last 5 years have seen over 100 publications per year.

These figures in fact can also be considered a little on the conservative side, since they do not include any research where Brønsted acids have been combined with metals or other catalysts, which would add to this number indefinitely. Since acid catalysis is perhaps one of the oldest forms of catalysis known and encompasses a wide spectrum of transformations, chiral Brønsted acids have found plentiful scope to feed upon. However, the catalysts over time have evolved as chemistry has progressed, and nowadays, they function as much more than a source of acidity, for example, they have been employed as chiral counterions, ligands, and phase-transfer agents, to name just a few. In addition, they also possess the ability to work in various mechanistic modes, which has added to their appeal and utility.

Current published literature generally focuses on the field of asymmetric organocatalysis, but this term encompasses such a variety of catalysts, mechanisms, and transformations, which usually forces the coverage to only realistically provide an overview of the surface of Brønsted acid catalysis. As part of our involvement in the field of asymmetric Brønsted acid catalysis, we have been monitoring and accumulating a vast number of publications within this area for the last 10 years. Given the scale of developments that have occurred, we sought it suitable to attempt to collate this wealth of information into a more manageable source of literature. This book therefore should serve as a timely essential to both novices and experts in the field. Furthermore, what differentiates this book over most review articles is that it will be structured primarily by structural motif and secondly by transformations. In doing so, the reader will be more aware of the chemical space that Brønsted acids occupy rather than an exhaustive list of reactions. We hope this book draws curiosity from all corners of organic chemistry, and we envision that it will be a platform for the development of future exciting research within the field, which is we look forward to see.

Aachen, Germany *Magnus Rueping, Dixit Parmar, Erli Sugiono*

1 Introduction

1.1 Book Structure and Notation

The field of asymmetric Brønsted acid catalysis is known to contain a diverse array of catalyst architectures, but by far the most dominant within these are BINOL-derived acids [1–11]. They have become so strongly associated with the term *asymmetric Brønsted acid* that it is not unusual to see this key descriptor omitted in abstracts of recent literature. To complicate the definition, several classes of hydrogen-bond donors such as BINOLs are sometimes referred to as *Brønsted acids*. While they may certainly be involved in hydrogen-bonding interactions, their complete dissociation to the conjugate base by loss of a proton is not a prevalent mechanistic pathway. The pKa values of common Brønsted acid catalysts have been measured by several groups [12–15] in a variety of solvents. Since the comparison of different acids is difficult across different solvents, we will not endeavor to formalize a pH range for the catalysts included in this book. Instead, in general, we have curtailed our coverage to only include BINOL-derived phosphoric acids (and derivatives), carboxylic acids, and sulfonamides.

In order to circumvent the repetitive reproduction of the most commonly employed catalysts within this book, we have opted to use a logical abbreviation formatting for the schemes (Figure 1.1).

BINOL-derived phosphoric acids will be defined as **PA**, the corresponding *N*-phosphoramides will be denoted as **NPA**, and finally, spirocyclic phosphoric acids will be referred to as **SPA**. All other catalysts will simply be denoted as **BA**. In each case, a number will follow the descriptor and these numbers will be consistent throughout the whole book. The nature of the aryl (or R-group) will be shown with the catalyst, and all **BA** catalysts will be embedded within the scheme. In most cases, the (*R*)-enantiomer is employed but wherever this differs, it will be noted appropriately, for example, **(*S*)-PA 1** would be the (*S*)-enantiomer. Finally, the partially dehydrogenated catalysts of BINOL will be referred to as $[H_8]$–, and this should be assumed as always being the distant aromatic rings.

Asymmetric Brønsted Acid Catalysis, First Edition. Magnus Rueping, Dixit Parmar and Erli Sugiono.

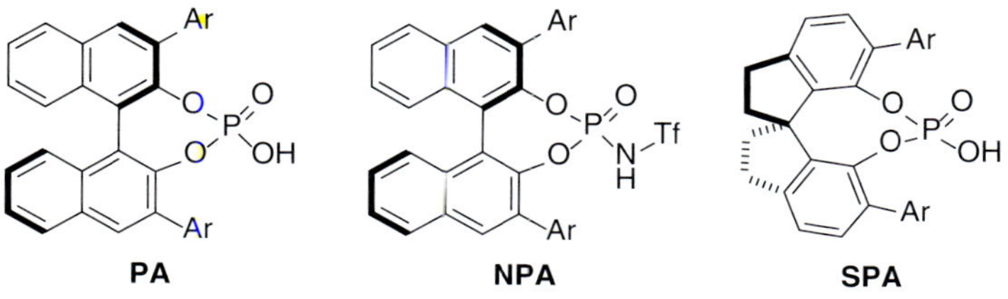

Figure 1.1 Commonly employed catalysts and their notation within this book.

1.2
Catalyst Preparation

This book will cover a range of Brønsted acid catalysts, so we will not endeavor to try and cover all the synthetic processes involved for every catalyst employed. However, since BINOL-derived phosphoric acids are featured in a large majority of research papers, we feel it would be beneficial to show the most commonly used route toward preparing them (Scheme 1.1) [16].

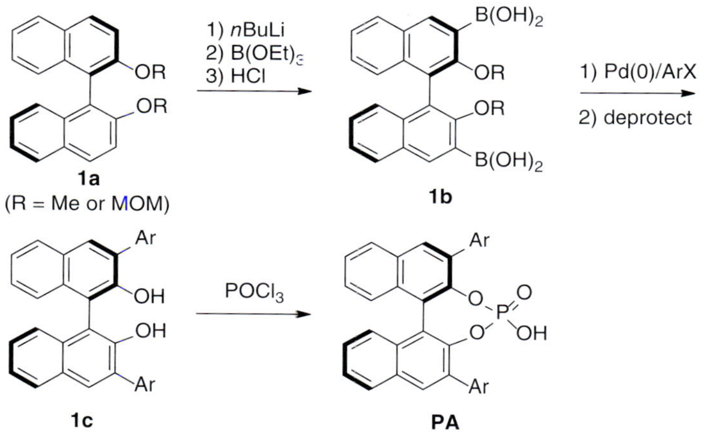

Scheme 1.1 Common synthetic route to BINOL phosphoric acid catalysts.

The process would normally commence from commercially available BINOL, which would be appropriately protected with either a methyl or a MOM group (**1a**). The next stage is to install boronic esters at the 3,3′-positions. This requires a three-step procedure, which involves lithiation, followed by electrophilic quenching with a boronate, and finally, hydrolysis to yield **1b.** The installation of aromatic groups is performed via a Suzuki reaction and then the protecting group can be removed to yield **1c**. Lastly, condensation of $POCl_3$ with BINOL **1c** leads after hydrolysis to the phosphoric acid catalyst **PA**. The overall route is highly reliable and works well for a range of aromatic partners. On a critical note, the route makes use of two wasteful protecting group steps, and to circumvent this flaw, several groups have reported protecting-group-free strategies [17, 18].

For the remaining common catalysts, the *N*-phosphoramide catalysts can be easily prepared in a single step from the corresponding phosphoric acid catalyst by a simple amidation reaction [19]. The spirocyclic phosphoric acid catalysts are prepared in a similar manner but starting from the corresponding spirobiindane diol precursor. This is, however, not commercially available and can be prepared as described by Birman *et al.* [20].

1.3 Metal Impurities

Although beyond the scope of this book, chiral phosphate anions combined with metal cations are well known to be suitable for catalyzing a whole host of asymmetric transformations [21–23]. Metal cations can enhance the catalytic behavior of a catalyst in a number of ways, and therefore, sometimes its role is difficult to determine. The issue of metal impurities is an even more difficult subject to address. Chiral phosphoric acids are now well known to chelate to metal cations during purification techniques, particularly so on silica gel. This usually leads to the isolation of a catalyst that contains a highly variable amount of metal content, which results in large deviations in catalytic performance from batch to batch. This phenomenon has been reported by several groups [24–28] and most likely has been experienced by many groups too. The solution to avoid such issues is simply to wash the purified catalyst with a strong acidic solution [25, 29]. Although researchers in the field are all conscious of this issue now, this was not always the case, and, therefore, it should be taken into account that some earlier reports may have involved metal impurities of which the authors were completely unaware . It is impossible to suggest which reports may suffer from this; therefore, we have adopted to include all the reports as reported originally.

References

1. Akiyama, T., Itoh, J., and Fuchibe, K. (2006) *Adv. Synth. Catal.*, **348**, 999–1010
2. Akiyama, T. (2007) *Chem. Rev.*, **107**, 5744–5758.
3. Terada, M. (2008) *Chem. Commun.*, **2008**, 4097–4112.
4. Terada, M. (2010) *Bull. Chem. Soc. Jpn.*, **83**, 101–119.
5. Terada, M. (2010) *Synthesis*, **2010**, 1929–1982.
6. Zamfir, A., Schenker, S., Freund, M., and Tsogoeva, S.B. (2010) *Org. Biomol. Chem.*, **8**, 5262–5276.
7. Kampen, D., Reisinger, C., and List, B. (2009) *Top. Curr. Chem.*, **291**, 395–456.
8. Terada, M. (2011) *Curr. Org. Chem.*, **15**, 2227–2256.
9. Parmar, D., Sugiono, E., Raja, S., and Rueping, M. (2014) *Chem. Rev.*, **114**, 9047–9153.
10. Rueping, M., Nachtsheim, B.J., Ieawsuwan, W., and Atodiresei, I. (2011) *Angew. Chem. Int. Ed.*, **50**, 6706–6720.
11. Cheon, C.H. and Yamamoto, H. (2011) *Chem. Commun.*, **47**, 3043–3056.
12. Christ, P., Lindsay, A.G., Vormittag, S.S., Neudörfl, J.-M., Berkessel, A., and O'Donoghue, A.C. (2011) *Chem. Eur. J.*, **17**, 8524–8528.

13. Yang, C., Xue, X.-S., Jin, J.-L., Li, X., and Cheng, J.-P. (2013) *J. Org. Chem.*, **78**, 7076–7085.
14. Kaupmees, K., Tolstoluzhsky, N., Raja, S., Rueping, M., and Leito, I. (2013) *Angew. Chem. Int. Ed.*, **52**, 11569–11572.
15. Yang, C., Xue, X.-S., Li, X., and Cheng, J.-P. (2014) *J. Org. Chem.*, **79**, 4340–4351.
16. Simonsen, K.B., Gothelf, K.V., and Jørgensen, K.A. (1998) *J. Org. Chem.*, **63**, 7536–7538.
17. Bartoszek, M., Beller, M., Deutsch, J., Klawonn, M., Köckritz, A., Nemati, N., and Pews-Davtyan, A. (2008) *Tetrahedron*, **64**, 1316–1322.
18. Li, B. and Chiu, P. (2011) *Eur. J. Org. Chem.*, **2011**, 3932–3937.
19. Nakashima, D. and Yamamoto, H. (2006) *J. Am. Chem. Soc.*, **128**, 9626–9627.
20. Birman, V.B., Rheingold, A.L., and Lam, K.-C. (1999) *Tetrahedron: Asymmetry*, **10**, 125–131.
21. Rueping, M., Koenigs, R.M., and Atodiresei, I. (2010) *Chem. Eur. J.*, **16**, 9350–9365.
22. Parra, A., Reboredo, S., Martín Castro, A.M., and Alemán, J. (2012) *Org. Biomol. Chem.*, **10**, 5001–5020.
23. Lv, J. and Luo, S. (2013) *Chem. Commun.*, **49**, 847–858.
24. Hatano, M., Moriyama, K., Maki, T., and Ishihara, K. (2010) *Angew. Chem. Int. Ed.*, **49**, 3823–3826.
25. Rueping, M., Nachtsheim, B.J., Koenigs, R.M., and Ieawsuwan, W. (2010) *Chem. Eur. J.*, **16**, 13116–13126.
26. Terada, M. and Kanomata, K. (2011) *Synlett*, (9), 1255–1258.
27. Della Sala, G. (2013) *Tetrahedron*, **69**, 50–56.
28. Enders, D., Stöckel, B.A., and Rembiak, A. (2014) *Chem. Commun.*, **50**, 4489–4491.
29. Klussmann, M., Ratjen, L., Hoffmann, S., Wakchaure, V., Goddard, R., and List, B. (2010) *Synlett*, **2010**, 2189–2192.

2
Reactions of Imines

2.1
Nucleophilic Addition Reactions

Nucleophilic addition reactions of imines are perhaps the most widely studied area within the chemical literature of chiral Brønsted acid catalysis. In fact, most transformations in this book would fall into this category, even though they may be known more formally by a named reaction. We have carefully chosen which specific reactions deserve to be grouped into their own category, and therefore in this section, we will cover the literature that does not fall discreetly in the upcoming subsections.

2.1.1
C-Nucleophiles

The construction of C–C bonds is an organic transformation that arguably has attracted the greatest level of attention from the synthetic community. This fundamental process is integral in the art of molecular assembly. Shortly following the seminal papers of chiral phosphoric acids by Akiyama and Terada as organocatalysts [1, 2], Terada developed the direct alkylation of *N*-acylimines **1b** with α-diazoesters **1a** catalyzed by **PA 1** (Scheme 2.1) [1].

The reaction with a variety of aromatic imines gave the corresponding products **1c** in good yields and high enantiomeric excesses. The nature of the protecting group was found to have a modest effect on the level of selectivity; while alkyl-protected imines failed to demonstrate any reactivity, *N*-acyl imines bearing a 4-$(NMe_2)C_6H_4$ substituent provided the highest selectivities. The products could be routinely transformed into β-amino acid derivatives without any loss of enantiopurity.

The mechanism of the process is depicted in Scheme 2.2. Activation of the imine by the catalyst is proposed to occur by hydrogen-bonding interactions with the basic nitrogen atom; however, such interactions may also be occurring with the oxygen atom of the amide unit. Addition of the diazoester to the imine occurs followed by an intramolecular deprotonation by the phosphoryl oxygen to yield the product. In 2012, Peng showed that α-diazophosphonates could also be used

Asymmetric Brønsted Acid Catalysis, First Edition. Magnus Rueping, Dixit Parmar and Erli Sugiono.

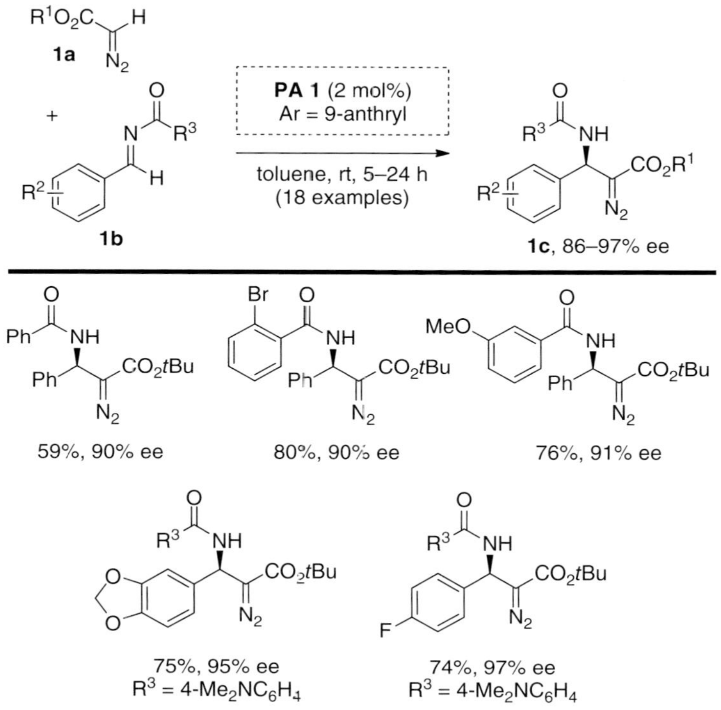

Scheme 2.1 Alkylation of α-diazoesters with imines by Terada [1].

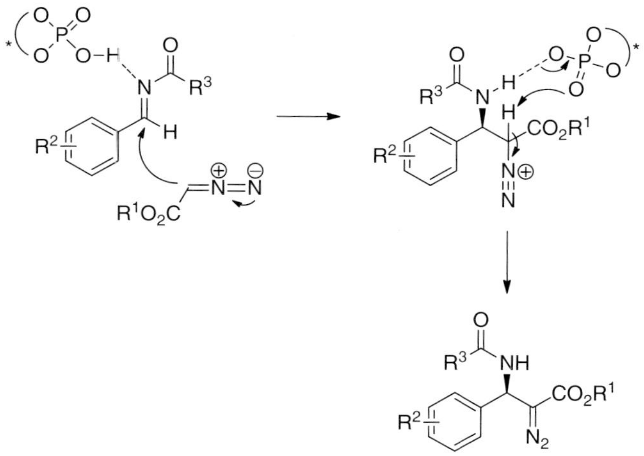

Scheme 2.2 Mechanism of diazoester additions to imines [1].

as suitable coupling partners to imines in the presence of only 0.1 mol% of a chiral phosphoric acid [3].

The field of chiral Brønsted acids is largely dominated by phosphoric acids and, surprisingly, carboxylic acids, which constitute an important class of acids, rarely feature as asymmetric catalysts. Over the past few years, Maruoka has been one of the pioneers attempting to bring carboxylic acid catalysts to the market. His seminal work reported on the addition of diazo-substrates **2a** to imines **2b** in the presence of chiral dicarboxylic acid **BA 1** (Scheme 2.3) [4, 5].

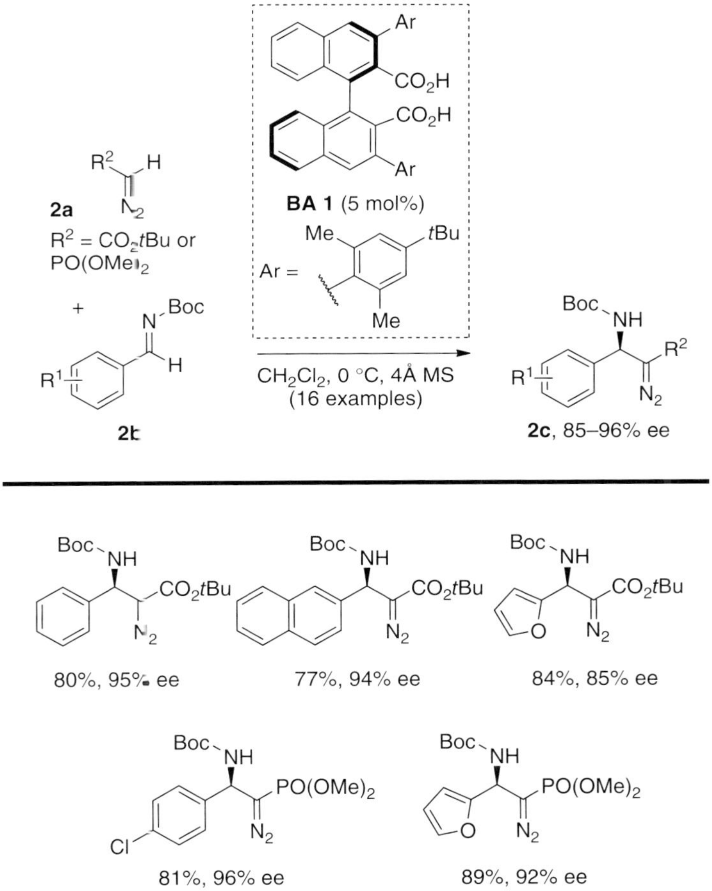

Scheme 2.3 Alkylation of diazoesters with imines by Maruoka [4].

The reaction proceeds under mild conditions and delivers the products **2c** in comparable yields and enantioselectivities to those reported by Terada (cf. Scheme 2.1) using a phosphoric acid. The structure of the catalyst has been elucidated, and it shows that it possesses a wide dihedral angle (93.4°) when compared to a typical phosphoric acid catalyst (55°) [6]. This property coupled with the presence of internal hydrogen bonding is thought to be crucial for the efficient control of stereoselectivity. The scope of this methodology has been extended by Maruoka to include diazosulfone and hydrazone components [6, 7].

The nature of the substituent of diazo-coupling partners is known to dictate the course of reactivity, and in 2008, Maruoka discovered that by using diazoacetamides **3a** with *N*-Boc imines **3b**, the reaction proceeded to yield *trans*-aziridine products **3c** in good yields and with high enantioselectivities (Scheme 2.4) [8]. Once again, a dicarboxylic acid (**BA 2**) was used, and this aza-Darzens reaction, as it is formally known, represents a useful expansion for the utility of this class of

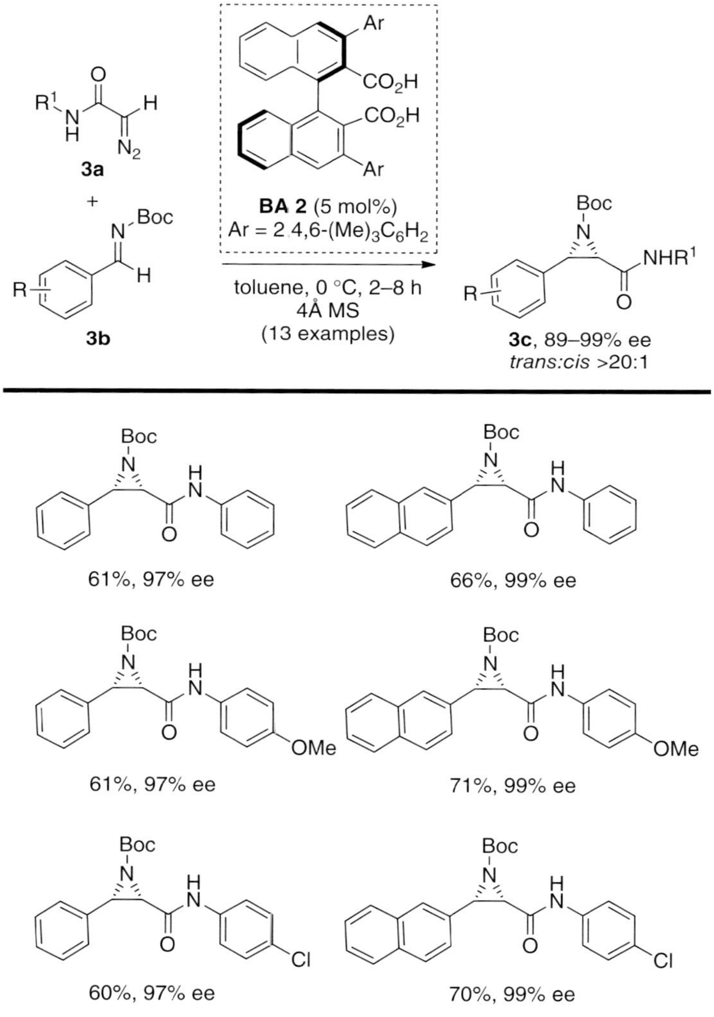

Scheme 2.4 Aziridination with imines by Maruoka [8].

catalysts. Maruoka has also performed this reaction with diazoacyl oxazolidinones, which provides access to tri-substituted aziridines [9].

A similar reaction was studied by Akiyama in 2009, who found that imines formed *in situ* from amine **3d** and glyoxals **3e** could react with diazoacetate **3f** to give *cis*-aziridines **3g** with high enantioselectivities (Scheme 2.5) [10]. Akiyama needed the use of just 2.5 mol% of **PA 2,** which interestingly provided the complementary diastereoisomer to Maruoka's work.

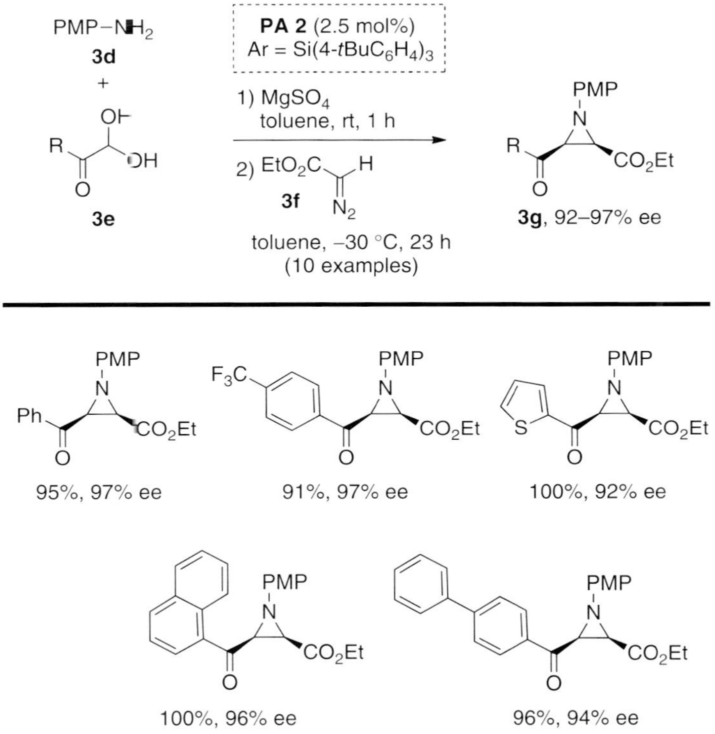

Scheme 2.5 Aziridination with imines by Akiyama [10].

The mechanism of aziridination is shown in Scheme 2.6. The process commences in a similar manner to the addition reaction whereby activation of the imine occurs followed by addition of the diazo-nucleophile. It is proposed that the acidity of the α-proton is reduced due to the adjacent amide group (as opposed to an ester group) and, therefore, deprotonation is less favorable. Instead, the nitrogen attacks the α-carbon, eliminating N_2 and forming the aziridine product. The acidity factor may well hold true for Maruoka's work, but Akiyama's aziridination process may be more reliant on a judicial choice of catalyst. This reaction has also received attention from Zhong, who utilized diazoacetamides with a phosphoric acid catalyst [11].

In 2006, Terada published an aza-ene-type reaction of enecarbamate **4a** with imines **4b** using a very low loading of 0.1 mol% of **PA 1** (Scheme 2.7) [12]. The

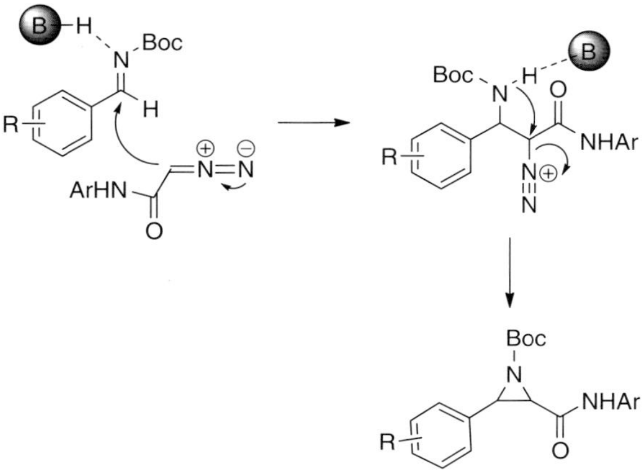

Scheme 2.6 Mechanism of aziridination with imines [8].

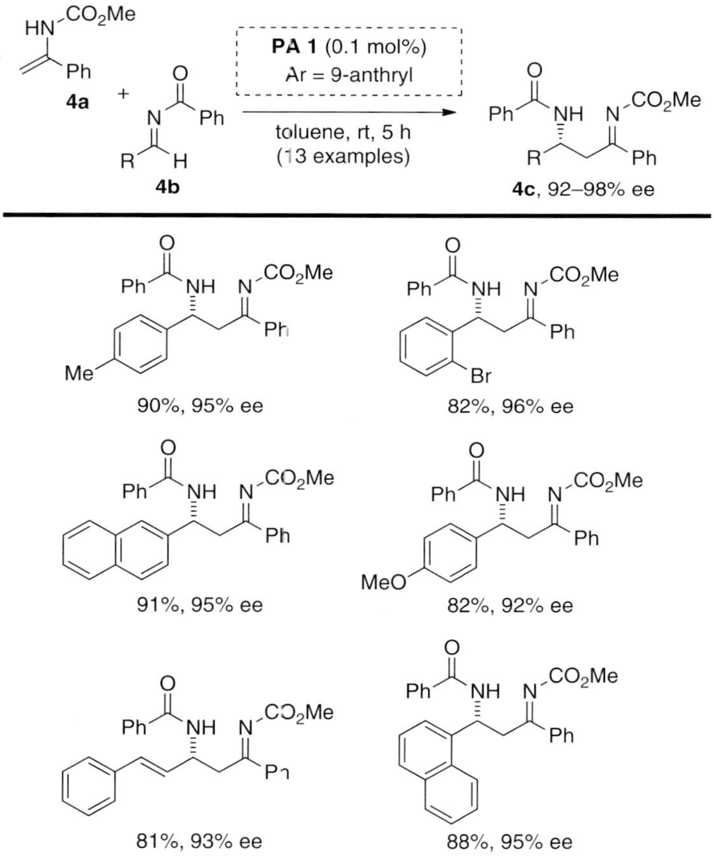

Scheme 2.7 Aza-ene reaction by Terada [12].

reaction efficiently yields **4c** with high enantioselectivities. Noteworthy is that the reaction can be performed on a gram scale while employing a substrate:catalyst ratio of 1000 : 1 without any detrimental performance.

The mechanism of the aza-ene reaction is outlined in Scheme 2.8. It is proposed that the phosphoric acid plays a bifunctional role during the transition, which involves simultaneous activation of both the imine and the enamine partners. Subsequent addition followed by proton transfer yields the desired product and regenerates the catalyst.

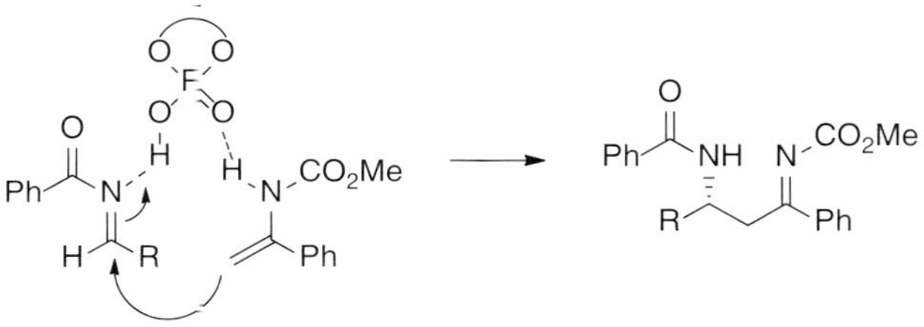

Scheme 2.8 Mechanism of aza-ene reaction [12].

Terada has explored the scope of this reaction and extended it to prepare piperidines by the use of 2 equiv. of enamine [13]. A multicomponent variant whereby the imine is generated *in situ* has also been explored by Masson and Zhu, which provides access to 1,3-diamines in good yields and with high enantioselectivities [14]. Mechanistically related Friedländer condensation, involving the addition of enamines to imines, has been shown by Gong to be suitably catalyzed by a phosphoric acid [15]. More recently, cascade processes involving *in situ* generated enamines and imines coupling together in an intramolecular fashion have been shown by Shi [16, 17].

In 2011, Momiyama and Terada disclosed the use of a fluorinated phosphoric acid catalyst (**[H$_8$]-PA 3**) to perform highly enantioselective Hosomi–Sakurai reactions of imines **5b** with allyl silane **5a** (Scheme 2.9) [18].

The reaction performs best when a stoichiometric amount of catalyst is used since the catalyst becomes silylated and hence is inactive after one cycle. It was, however, shown that, when a racemic phosphoric acid is used in conjunction with 20 mol% of **[H$_8$]-PA 3**, satisfactory enantioselectivities of the products **5c** could be achieved. List has also studied this reaction with disulfonimide catalyst **BA 3** to deliver similar products **5g** with good enantioselectivities (Scheme 2.10) [19].

The mechanism of the allylation is illustrated in Scheme 2.11. Activation of the imine occurs by hydrogen-bonding interactions with the phosphoric acid catalyst. This is followed by the attack by the allyl silane to form a β-silyl carbocation, which is captured by the phosphate anion of the catalyst. This leads to the products

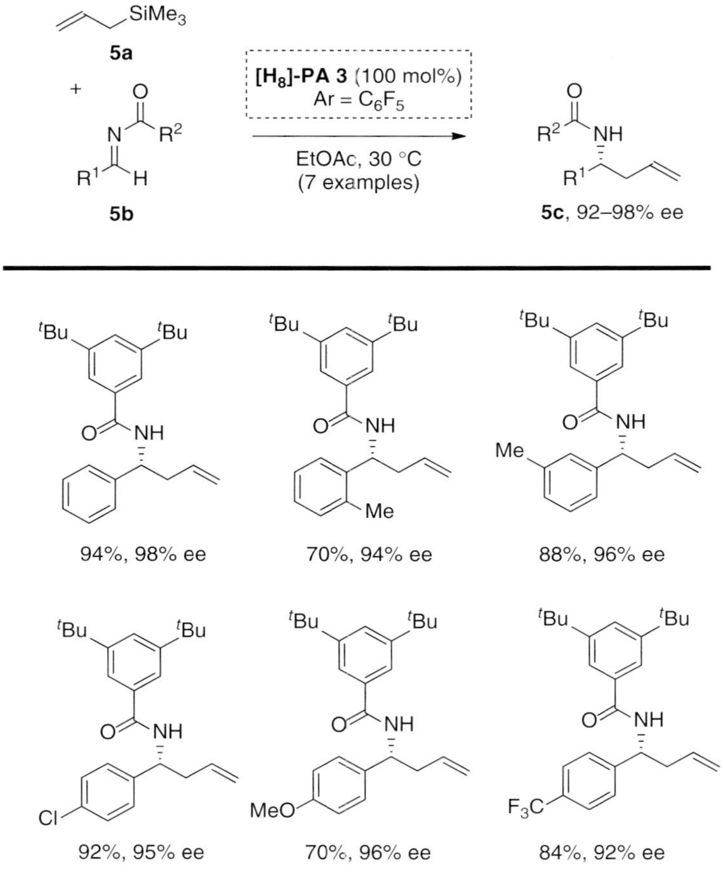

Scheme 2.9 Allylation of imines by Momiyama and Terada [18].

and the silylated catalyst. In the nonstoichiometric variant, an achiral additive (usually an acid) would desilylate the catalyst and thus the process would become catalytic.

Hydrazones are recognized for their nucleophilicity due to the resonance effect of the nitrogen's lone pair and serve as highly versatile synthetic intermediates. In 2007, Rueping demonstrated the coupling of hydrazones **6b** with imines **6a** in the presence of catalyst **[H$_8$]-PA 4** (Scheme 2.12) [20]. These mild conditions deliver the valuable products **6c** with good selectivities.

A similar approach has been described by Maruoka (Scheme 2.13), who showed the use of catalyst **BA 1** to yield **6f** with comparable enantiopurities [21]. Interestingly, although both groups employed the (*R*)-isomer of catalyst, complementary enantiomers were produced, suggesting once again that the mode of action may well differ between these catalysts.

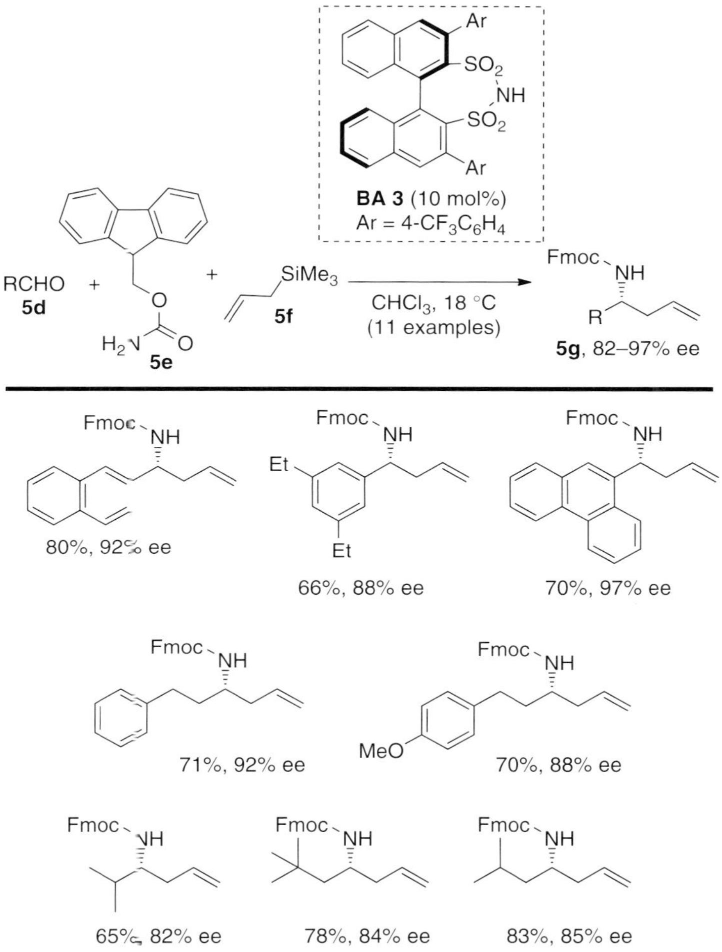

Scheme 2.10 Allylation of imines by List [19].

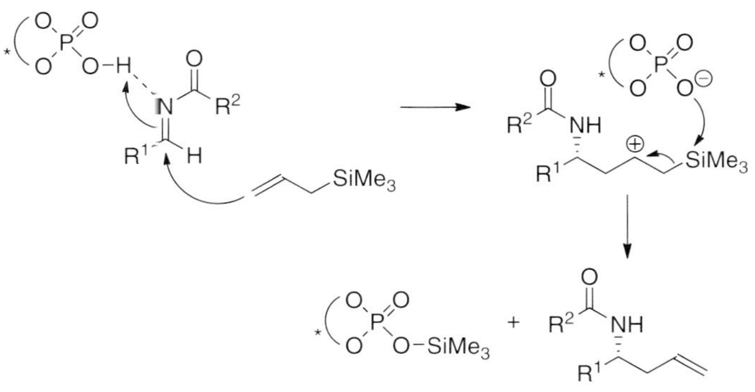

Scheme 2.11 Mechanism of allylation of imines [18].

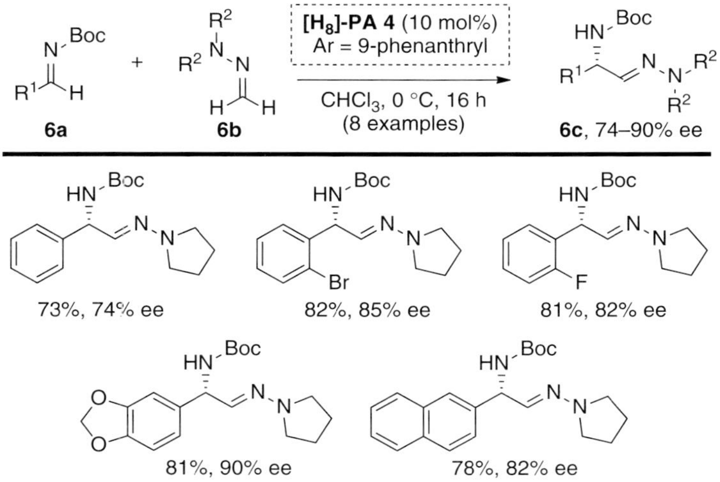

Scheme 2.12 Hydrazone addition to imines by Rueping [20].

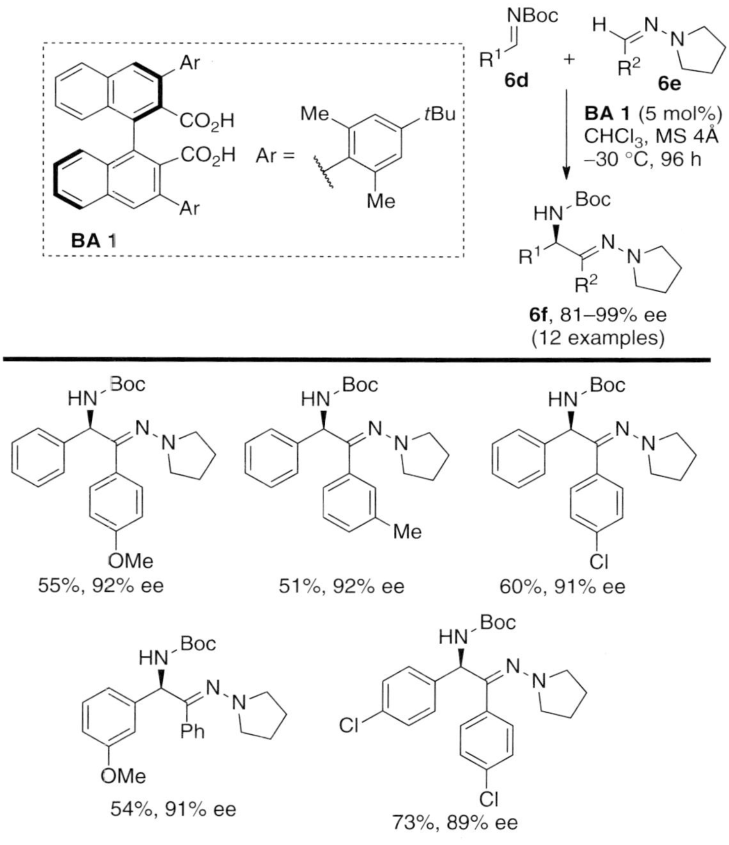

Scheme 2.13 Hydrazone addition to imines by Maruoka [21].

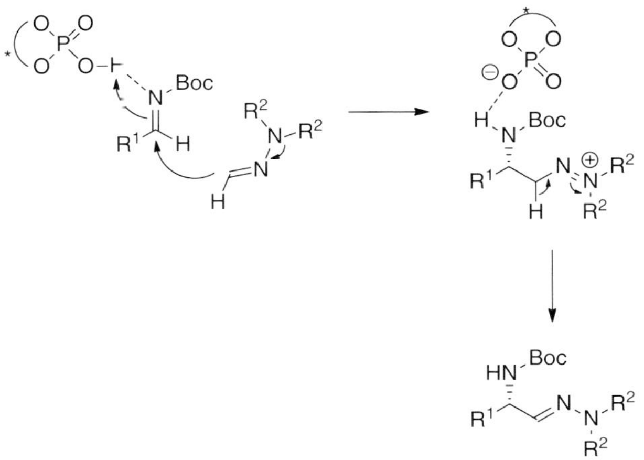

Scheme 2.14 Mechanism of hydrazone addition to imines by Rueping [20].

The mechanism of hydrazone addition into imines is detailed in Scheme 2.14. Activation of the imine occurs by the catalyst followed by addition of the hydrazone. This generates a cationic intermediate, which must lose a proton to yield the desired products and regenerate the catalyst. Maruoka has also studied the vinylogous variant of this reaction [22].

The Ugi reaction represents an ideal protocol for generating molecular diversity by means of varying any one of the four components involved in the reaction. The key step involved in the reaction is the nucleophilic addition of an isocyanide to an imine, which is a concept used by Wang and Zhu to form aminooxazoles **7d** from **7a**, **7b**, and **7c** (Scheme 2.15) [23].

A relatively high loading of **PA 5** was needed to effect modest selectivity, and it was suggested that this was due to the undesired interactions of the catalyst with the products. Masson and Zhu have furthered progress on this reaction to include four components to access enantioenriched polyheterocycles [24].

The mechanism of the addition of isocyanides is shown in Scheme 2.16. Firstly, the aldehyde and amine components can condense to form an imine intermediate. This imine is then activated by the acid catalyst, which allows the nucleophilic isocyanide to attack. The generated carbocation can be intercepted by the amide, and following a proton transfer, the desired aminooxazoles are formed in addition to regeneration of the catalyst. Additionally, Maruoka has also studied isocyanide addition to hydrazones using a dicarboxylic catalyst [25]. A selection of other carbon-based nucleophiles has also been demonstrated in various transformations with imine coupling partners. These include cyclic enol ethers [26–28] and nitroalkanes [29].

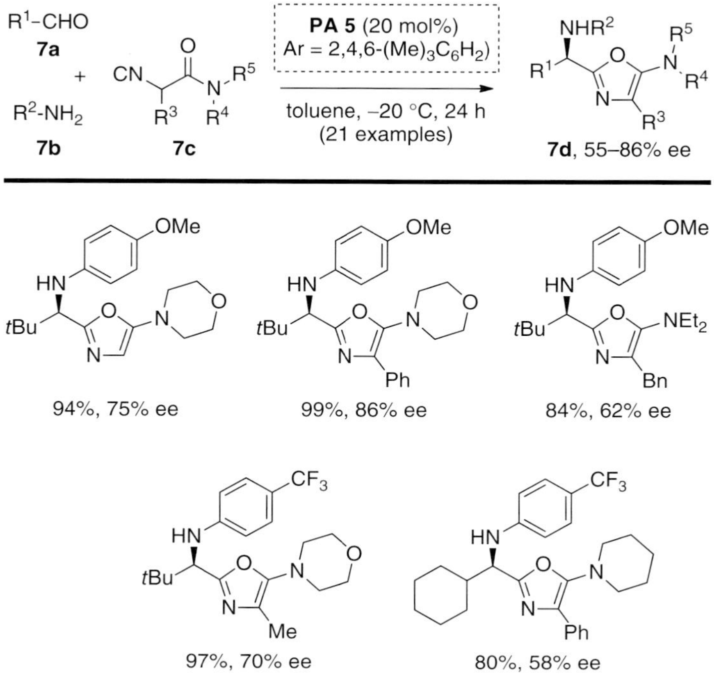

Scheme 2.15 Addition of isocyanides to imines by Wang and Zhu [23].

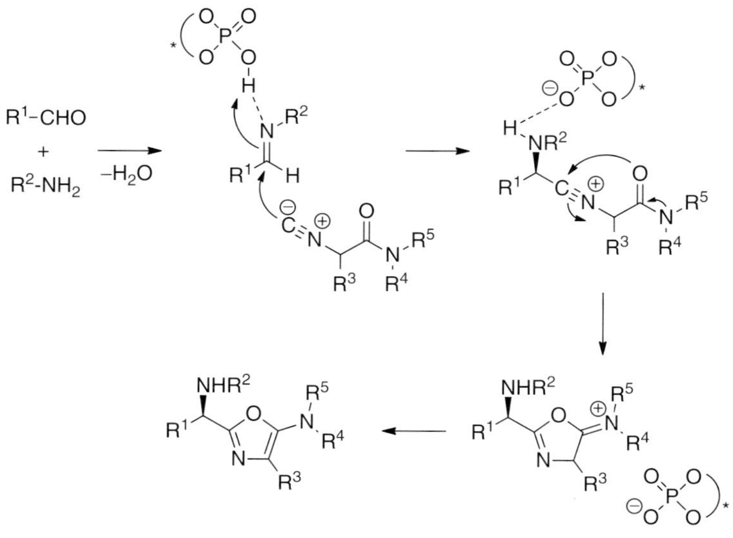

Scheme 2.16 Mechanism of isocyanide addition to imines [23].

2.1.2
N-Nucleophiles

The addition of nitrogen nucleophiles to imines has received much less attention than expected due to the inherent difficulty in achieving a successful reaction. The major obstacle is finding a nucleophile that is not basic enough to be protonated by the Brønsted acid but sufficiently reactive enough to perform the addition reaction. In 2005, Antilla was the first to recognize that sulfonamides **8a** could be used to react with acyl-imines **8b** to yield the *N*,*N*-aminals **8c** with high levels of enantioselectivity (Scheme 2.17) [30].

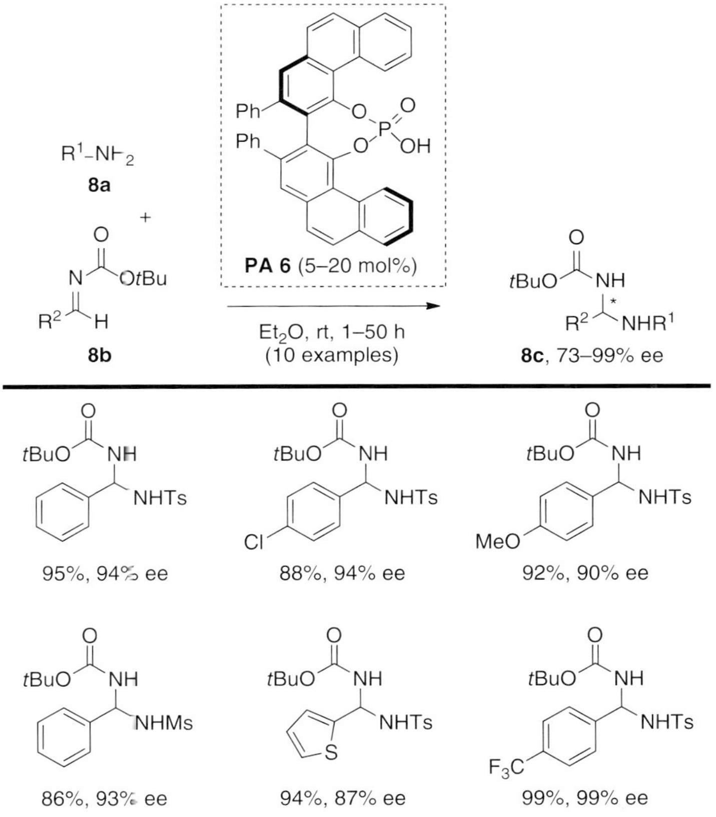

Scheme 2.17 Intermolecular addition of amines to imines by Antilla [30].

A screening of BINOL-derived phosphoric acids gave moderate results, but the employment of a VAPOL-derived phosphoric acid **PA 6** achieved higher selectivities albeit with a high loading in some cases. Antilla has extended the process to imide nucleophiles [31], and Ishihara has employed a chiral ammonium disulfonate catalyst in a similar procedure [32].

The intramolecular variant of this process was realized a few years later almost simultaneously by the groups of List and Rueping. List developed the use of aliphatic aldehydes **9a** as condensation partners for 2-aminobenzamides **9b** to access dihydroquinazolinones **9c** (Scheme 2.18) [33]. It was found that heavily hindered catalyst **(*S*)-PA 7** was crucial to achieving acceptable levels of enantioselectivity.

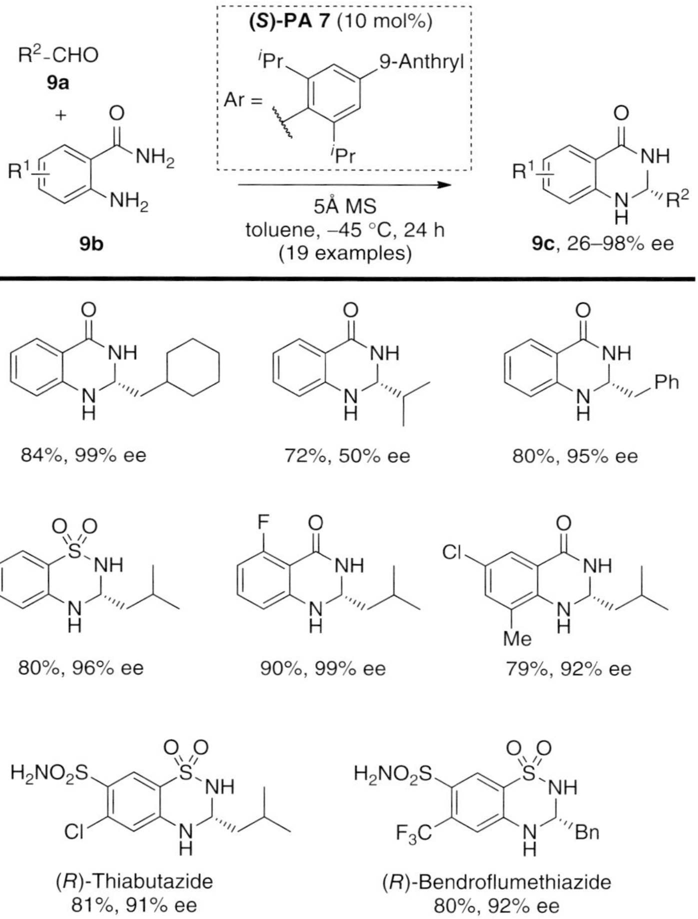

Scheme 2.18 Condensation of aldehydes with N-nucleophiles by List [33].

Rueping developed his methodology with aromatic aldehydes **9d** and benzamide **9e** using more commonly employed **PA 1** as the catalyst to access **9f** with good selectivities (Scheme 2.19) [34]. A representative mechanism is portrayed in Scheme 2.20. Condensation of the aldehyde with the aniline initially leads to an imine precursor. This can be suitably activated by the acidic proton of the phosphoric acid catalyst while the phosphoryl oxygen can interact with protons located

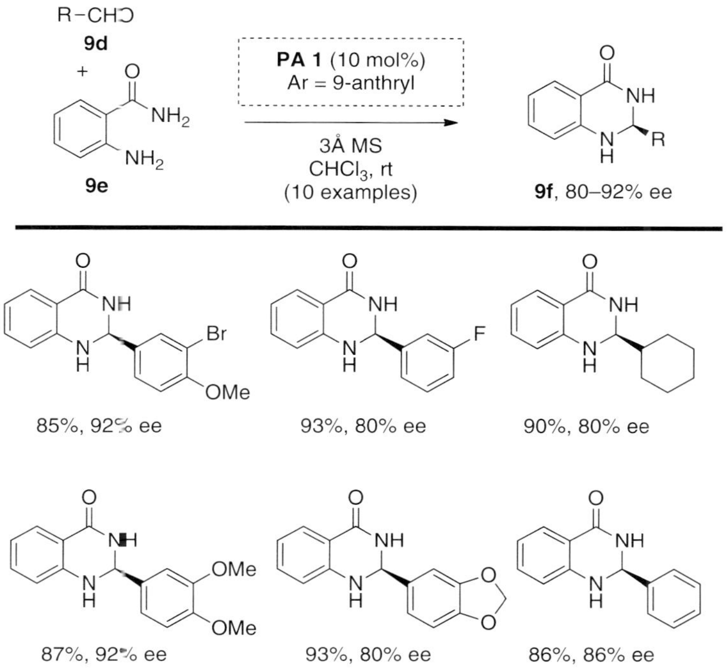

Scheme 2.19 Condensation of aldehydes with N-nucleophiles by Rueping [34].

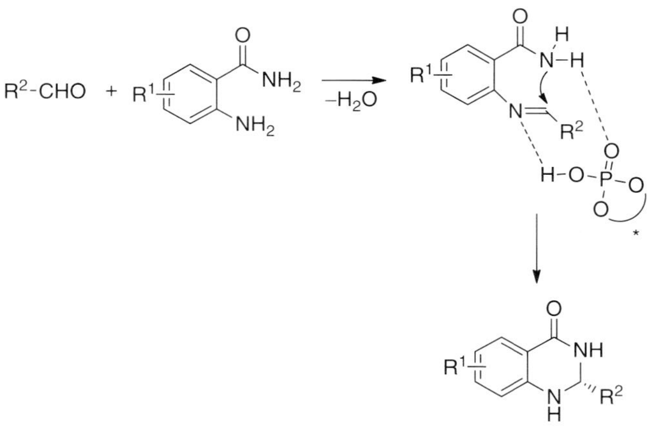

Scheme 2.20 Mechanism of the condensation reaction.

on the amide nitrogen. Subsequent intramolecular attack leads to the product after a simple proton transfer step.

Several related reports concerning this transformation have since appeared, which include the use of imines [35, 36] or isatins [37] as coupling partners and the employment of a SPINOL-derived catalyst [38].

2.1.3 O-Nucleophiles

The asymmetric addition of O-nucleophiles to activated imines is a challenging prospect since the products formed usually are not stable (due to the ease of collapsing an *N*,*O*-aminal) and the addition proceeds in the absence of a catalyst. In 2008, Antilla was able to show that simple alcohols **10a** could be coupled with acyl imines **10b** to give *N*,*O*-aminals **10c** with good enantioselectivities (Scheme 2.21) [39].

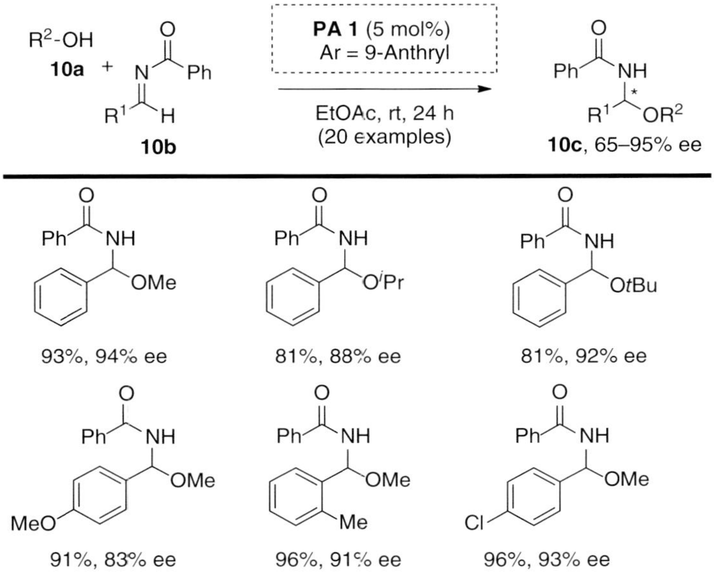

Scheme 2.21 Addition of alcohols to imines by Antilla [39].

The scope of alcohol partners ranged from simple MeOH to more complex long-chain alcohols containing various functionalities. Interestingly, catalyst **PA 1** provides the highest selectivities when performed at room temperature (rather than at lower temperatures) and with the use of polar solvents. Antilla has extended this methodology to also include peroxide [40] and thiol [41] nucleophiles. In a closely related procedure, the addition of alcohols to imines to perform dynamic kinetic resolutions of azlactones has been shown by the groups of Birman [42] and Gong [43].

An extension of the formation of *N*,*N*-aminals (cf. Scheme 2.18) was presented by List in 2010 by showing the use of 2-hydroxybenzamides **11a** as coupling partners for various aldehydes **11b** to yield the corresponding cyclic *N*,*O*-aminals **11c** (Scheme 2.22) [44].

The optimization of a suitable catalyst proved to be challenging, as standard phosphoric and *N*-triflyl phosphoramide catalysts provided only low levels of

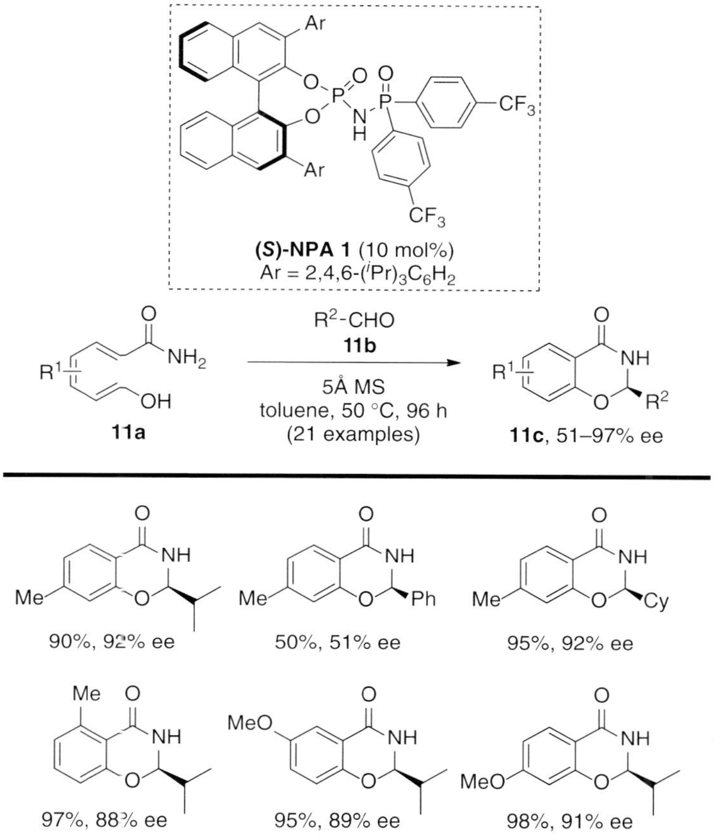

Scheme 2.22 Condensation of aldehydes with N- and O-nucleophiles by List [44].

control. Instead, the group choose to design a novel bisphosphorylimide catalyst ((*S*)-**NPA 1**), which contained aryl groups on the phosphinyl moiety. It was found that these groups exerted a considerable effect on the enantioselectivity and the authors propose that the basic phosphinyl oxygen activates the nucleophile in the transition state.

2.1.4 P-Nucleophiles

Although the use of P-centered nucleophiles is perhaps not frequently encountered, they represent a significant area of research considering the importance of chiral phosphorus-containing compounds. Akiyama published the seminal paper in 2005 regarding the addition of phosphite to imines; however, only modest ee's could be obtained [45]. In 2008, List reported the Kabachnik–Fields reaction to

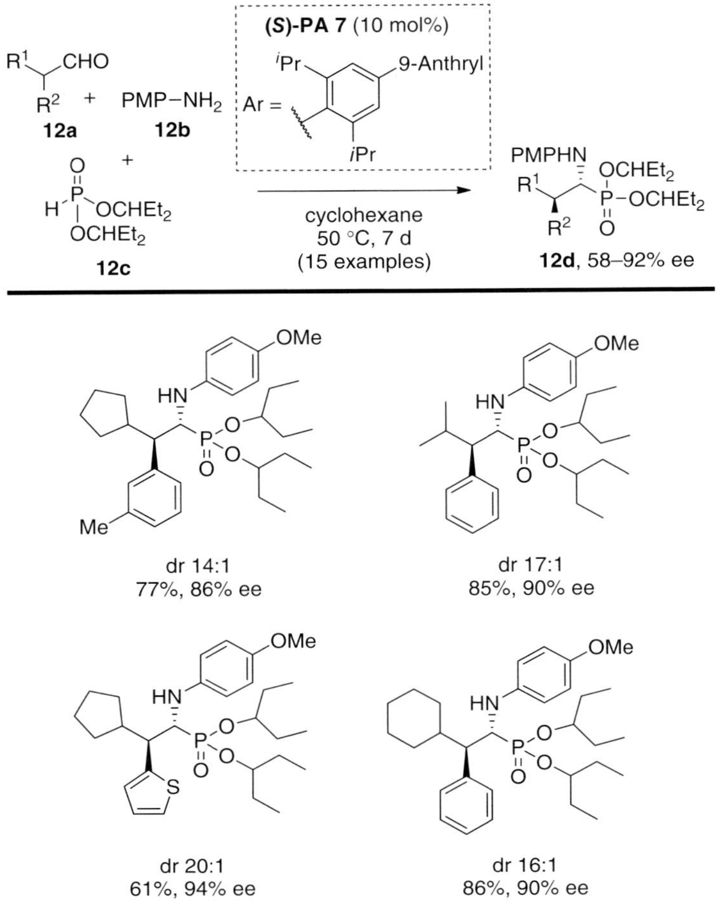

Scheme 2.23 Kabachnik–Fields reaction by List [46].

access α-aminophosphonates **12d** (Scheme 2.23) [46]. The reaction, which is formally a dynamic kinetic resolution of **12a**, combines **12a**, aryl amine **12b**, and phosphite **12c** in the presence of sterically encumbered catalyst **(*S*)-PA 7**.

The impressive protocol fuses a dynamic kinetic resolution with the formation of an additional stereocenter to yield **12d** with high enantioselectivities. Bhusare has also developed a proline-derived organocatalyst (**BA 4**), which has been shown to be proficient in catalyzing the addition of phosphites to imines to yield **12h** with high selectivities (Scheme 2.24) [47].

The mechanism is illustrated in Scheme 2.25. After condensation of the aldehyde and amine components, an imine intermediate is formed. This can be activated by the catalyst along with the phosphite nucleophile. This tautomer is the reactive form; however, the equilibrium lies very much toward the unreactive phosphonate tautomer. After subsequent attack and proton transfer, the desired

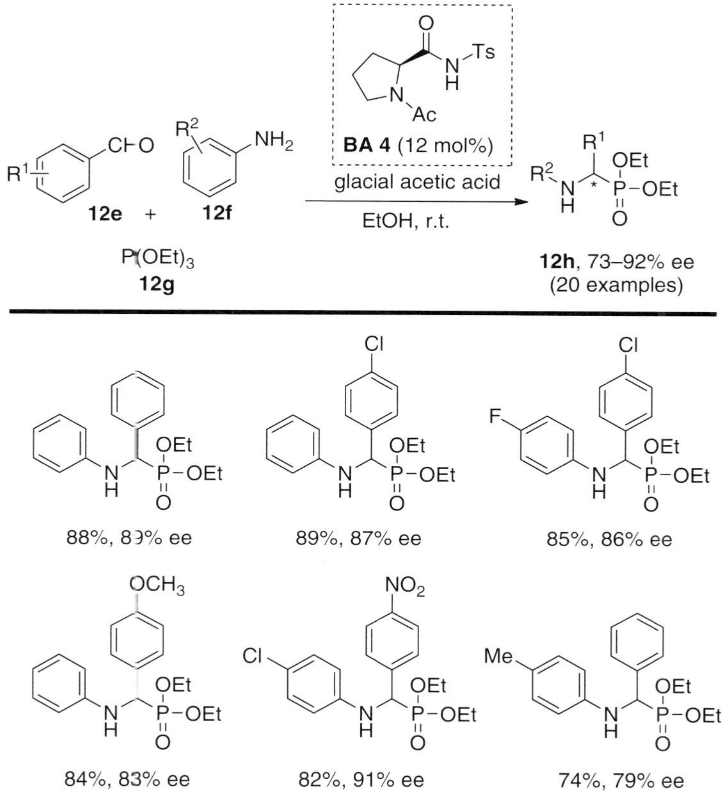

Scheme 2.24 Kabachnik–Fields reaction by Bhusare [47].

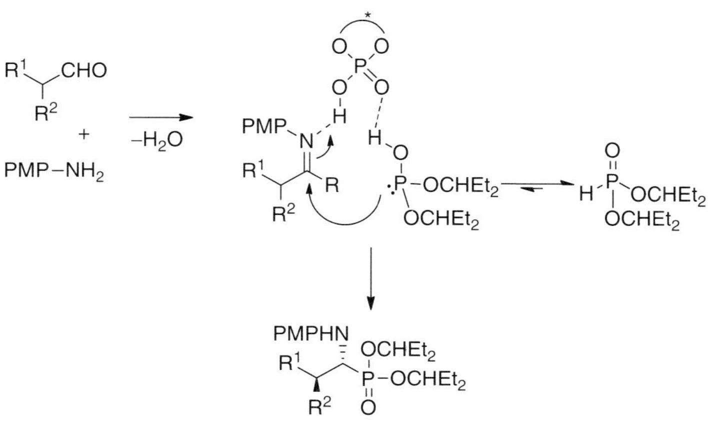

Scheme 2.25 Mechanism of Kabachnik–Fields reaction [46].

product is formed. Related transformations have been studied by the groups of Lin [48] and Song [49] using chiral phosphoric acids.

2.2 Mannich Reactions

The Mannich reaction is now considered as the benchmark asymmetric transformation, which can be catalyzed by a plethora of organocatalysts. It is frequently used by research groups as the testing ground in the development of new catalysts. It is also famously known as the *protocol,* which kick-started the field of chiral Brønsted acids. The seminal paper in this field was disclosed by Akiyama in 2004, who developed the coupling of silyl ketene acetals **13b** with 2-hydroxy phenyl protected imines **13a** to give the Mannich products **13c** with high enantiopurities (Scheme 2.26) [2, 50].

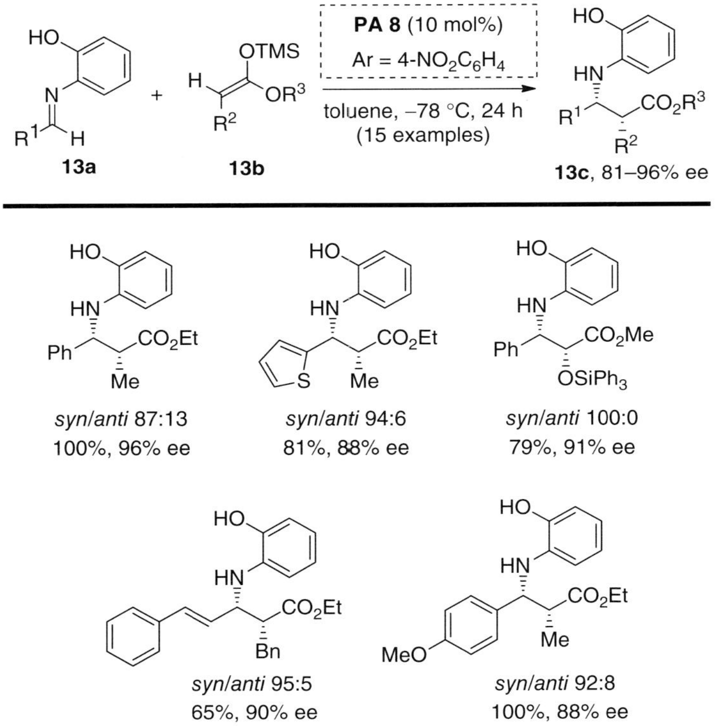

Scheme 2.26 Seminal reports using chiral phosphoric acids for the Mannich reaction by Akiyama [2].

From the selection of catalysts tested, it was clear that steric hindrance was not the only factor contributing to the enantioselectivity, since sterically more hindered catalysts than **PA 8** did not result in superior enantioselectivities.

In addition, it was recognized here and later confirmed by DFT calculations [51] that the 2-hydroxy functionality was crucial to the level of selectivity obtained. Later that year, Terada reported a direct Mannich reaction catalyzed by a chiral BINOL-derived phosphoric acid between imines **13d** and diketone **13e** to give the corresponding products **13f** with excellent enantioselectivities (Scheme 2.27) [52]. In contrast to Akiyama, sterically more hindered catalysts performed better and mechanistic studies by the Goodman group [53] have elucidated it to be proceeding via an alternative hydrogen-bonding pathway.

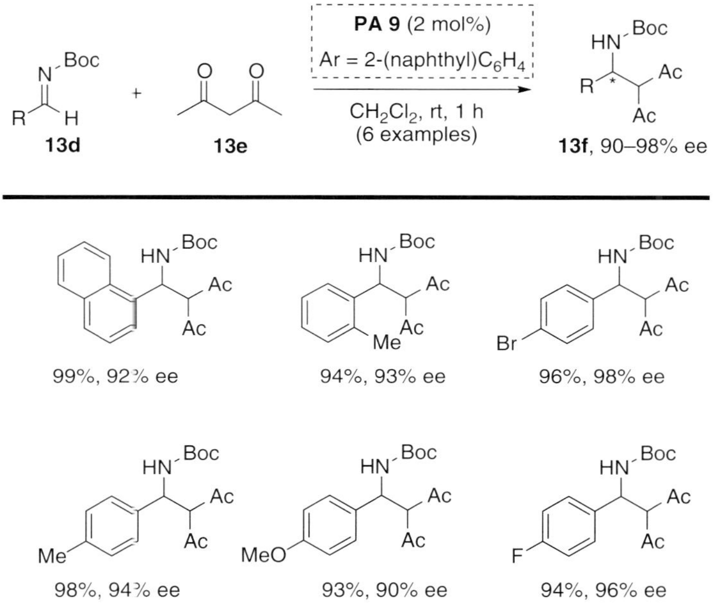

Scheme 2.27 Seminal reports using chiral phosphoric acids for the Mannich reaction by Terada [52].

The mechanism of Akiyama's approach is shown in Scheme 2.28. The phosphoric acid catalyst activates the 2-hydroxyphenyl protected imine by two discreet hydrogen-bonding interactions. This allows the silyl ketene acetal to attack, and loss of the TMS group is facilitated by the phenol, which allows for the catalyst to regenerate. Upon an aqueous workup, the desired product is isolated. Closely related transformations have also been reported by Gong [54], Rueping [55], and Amarante [56], who employ chiral phosphoric acids, and by Ishihara, who employs a chiral disulfonamide [57].

As mentioned at the start of this chapter, the Mannich reaction has attracted the attention of many research groups as a development hot spot for the design of new catalysts. Terada was one of the first who showed a phosphorodiamidic acid catalyst, but only modest enantioselectivities were obtained [58]. Several groups have had more success, and some of the best catalysts are shown in Figure 2.1.

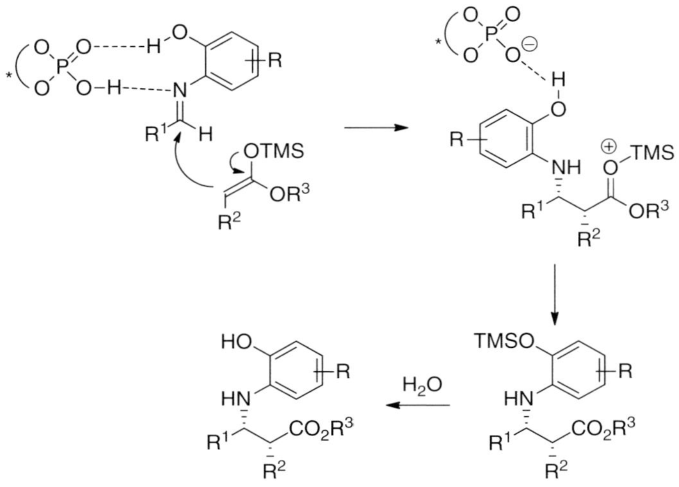

Scheme 2.28 Mechanism of the Mannich reaction [2].

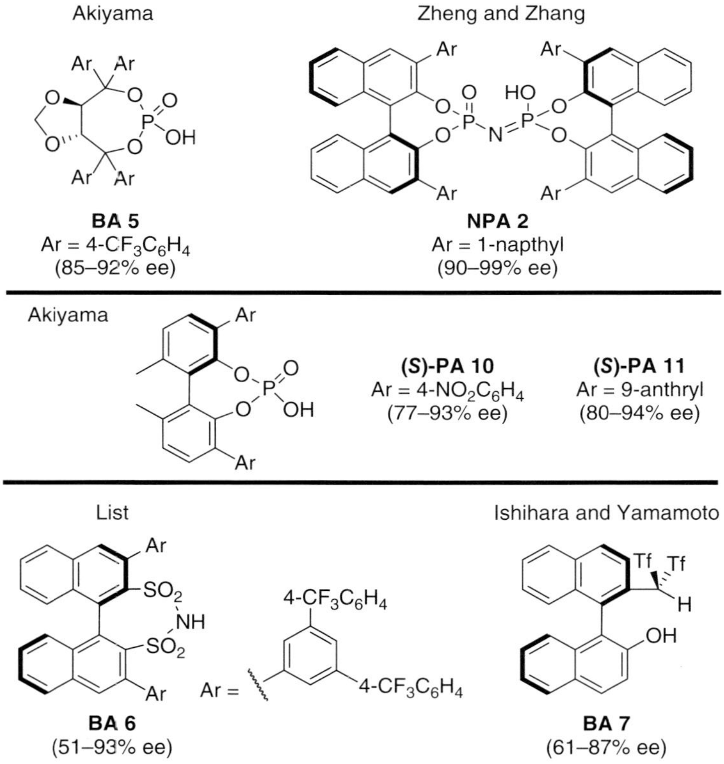

Figure 2.1 Chiral acid catalysts used for the Mannich reaction.

Akiyama has been particularly prevalent in this area by the progression of a catalyst derived from TADDOL (**BA 5**) [59] and by simplified catalysts based on a biphenyl core ((*S*)-**PA 10** and **11**) [60, 61]. Zheng and Zhang have shown a doubly axially chiral bisphosphorylimide catalyst **NPA 2** that is able to achieve excellent selectivities [62]. Finally, the groups of List [63] and Ishihara and Yamamoto [64] have developed disulfonimide **BA 6** and BINOL-like catalyst **BA7**, respectively, toward performing the Mannich reaction of imines with silyl ketene acetals.

The closely related vinylogous Mannich reaction is a useful extension of the regular variant as it provides highly valuable synthetic intermediates suitable for further transformations. In 2008, Schneider developed this reaction by reacting aryl-protected imines **14a** with vinylogous ketene acetal **14b** in the presence of **PA 5** (Scheme 2.29) [65].

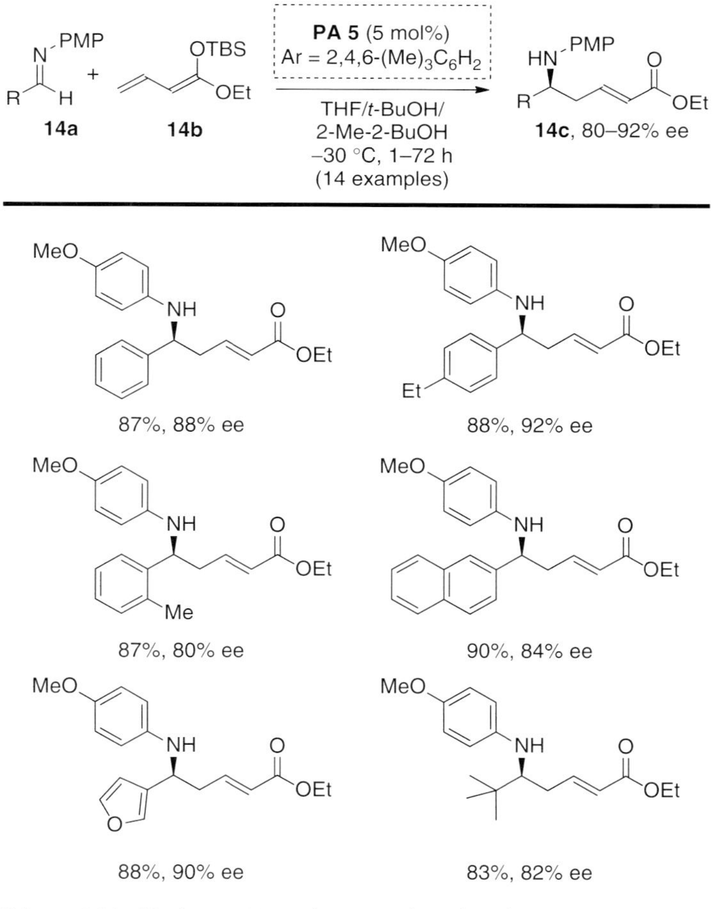

Scheme 2.29 Vinylogous Mannich reaction by Schneider [65].

The group found that coordinating solvents greatly improved the reaction rate and the enantioselectivity of **14c**, and after optimization, a mixture consisting of equal amounts of THF, *t*BuOH, and 2-Me-2-BuOH with the addition of 1 equiv. of water was found to be the optimal medium. The Schneider group has been influential in this area and has published several extensions to this methodology [66–69], which includes the synthesis of a tobacco alkaloid [70]. Asymmetric vinylogous Mannich reactions catalyzed by chiral phosphoric acids have also been studied by other research groups for various applications [71–74].

2.3 Strecker Reactions

The Strecker reaction is a practical route to synthesize amino acid derivatives from the correspondingly produced α-amino nitriles. The combination of simple starting materials allows for the formation of natural and unnatural derivatives. In 2006, Rueping published an asymmetric protocol, which was effectively catalyzed by **PA 4**. The reaction of benzyl protected imines **15a** with hydrogen cyanide **15b** gave the Strecker products with high enantioselectivities (Scheme 2.30) [75, 76].

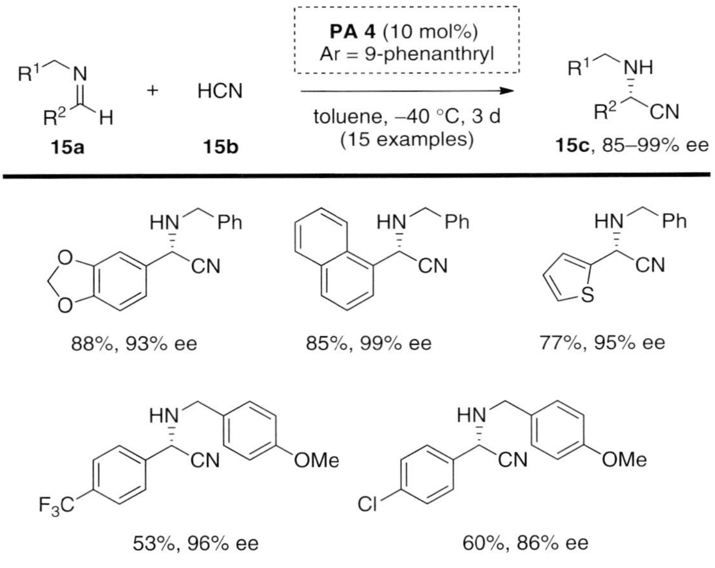

Scheme 2.30 Strecker reaction by Rueping [75].

Recently, Dughera has developed a disulfonimide catalyst ((−)-**BA 8**), which has been shown to deliver Strecker products **15g** with good selectivities (Scheme 2.31) [77]. The mechanism of the process has been studied in-depth by Goodman and elucidated to consist of hydrogen-bonding interactions of both components

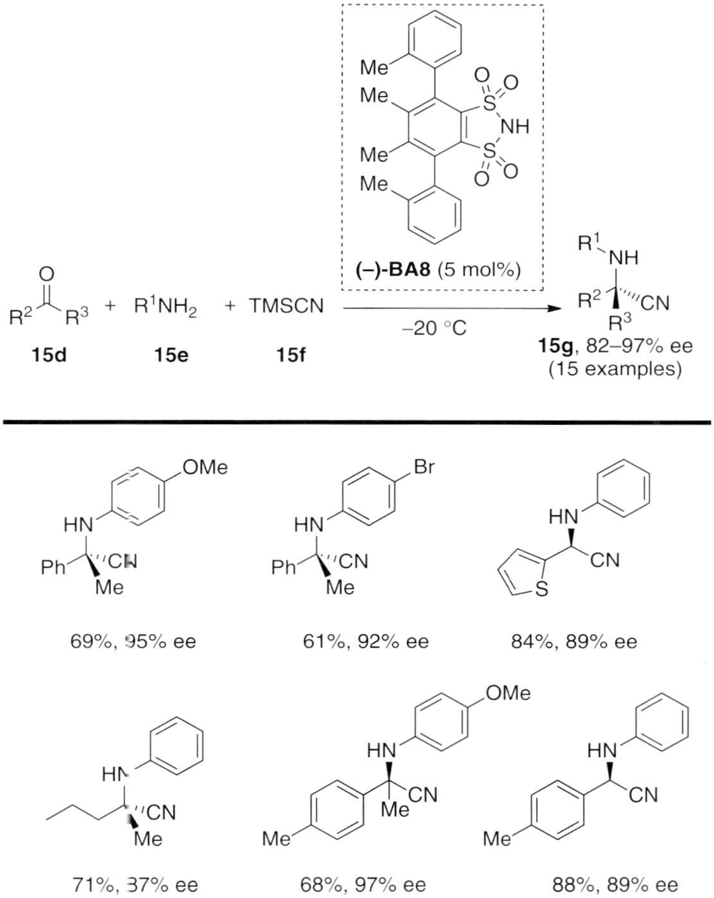

Scheme 2.31 Strecker reaction by Dughera [77].

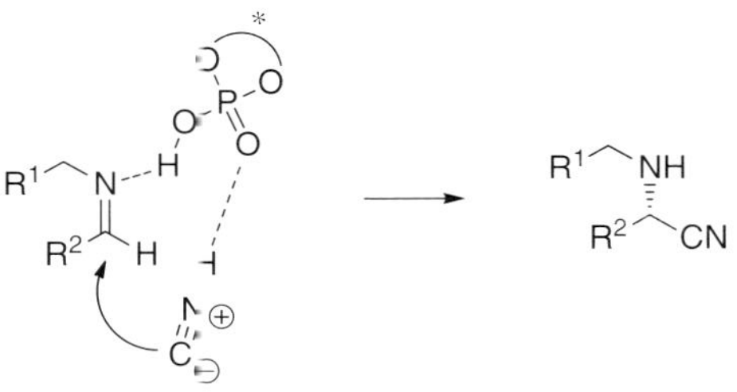

Scheme 2.32 Mechanism of the Strecker reaction [75].

to the catalyst [78]. A snapshot of the mechanism is shown in Scheme 2.32. As predicted by computational calculations, the catalyst is involved in hydrogen bonding to both the imine and cyanide moieties. Closely related works by the groups of Tsogoeva [79] and Ma [80] have also been reported using phosphoric acid catalysts.

The major drawback of the Strecker reaction lies in the handling of toxic cyanide sources, which must be handled with extreme precautions. Maruoka presented an alternative route to Strecker products via an asymmetric imino aza-enamine reaction (Scheme 2.33) [81].

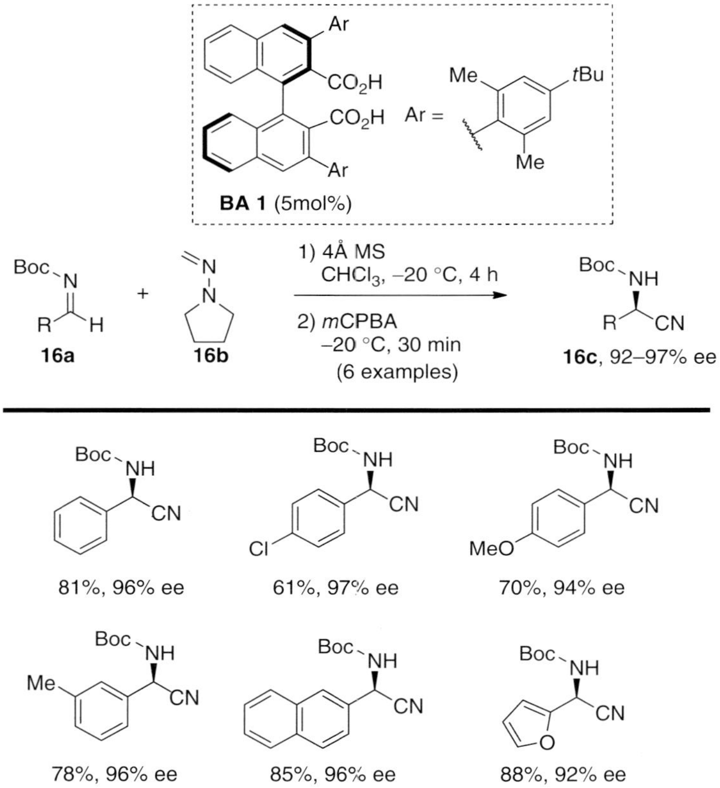

Scheme 2.33 Strecker-type reaction by Maruoka [81].

The couplings of Boc-protected imines **16a** with hydrazone **16b** lead to **16c** with excellent enantioselectivities. The reaction is catalyzed by axially chiral dicarboxylic acid **BA 1**, which could be lowered to a loading of 0.1 mol% without any detrimental effects.

The mechanism of the Strecker-type reaction is depicted in Scheme 2.34. Activation of the imine proceeds by the catalyst while attack by the imino aza-enamine occurs. Following a proton transfer, the catalyst is reformed and an intermediate hydrazone is produced. Upon oxidation by *m*CPBA, the corresponding Strecker product can be isolated.

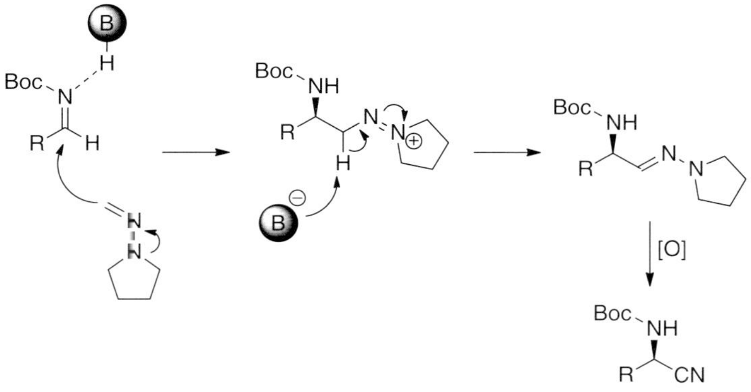

Scheme 2.34 Mechanism of the Strecker-type reaction [81].

2.4 Biginelli Reactions

Multicomponent reactions are attractive protocols for chemists who wish to synthesize a library of similar compounds with various points of molecular diversity. Within this class, the Biginelli reaction holds a noteworthy position as it offers an entry into nitrogen heterocycles, which are of importance due to displayed pharmacological activities. The seminal report of a variant of the Biginelli reaction came from Gong in 2006, who showed that combination of **17a**, **17c**, and ureas **17b** gave enantioenriched dihydropyridines **17d** with good to high selectivities (Scheme 2.35) [82].

Interestingly, larger substituents at the 3- and 3′-positions of the catalyst gave lower yields and enantioselectivities of the products. In the end, the optimal catalyst was found to be simply **[H$_8$]-PA 12**.

The mechanism of the process is portrayed in Scheme 2.36. Condensation of the aldehyde and urea components leads to an imine-like intermediate. This can be subsequently activated by the catalyst while the enol ether proceeds to attack the carbon. Upon generation of a second urea intermediate, internal condensation between the amine and carbonyl groups leads to the product. Three years later, Gong found that in fact the absolute stereochemistry of the products could be tuned by the employment of different groups attached to the 3- and 3′-positions of the catalyst [83]. An illustrative example is shown in Scheme 2.37.

Combination of **18a**, **18b**, and **18c** gave in the presence of **PA 13**, the *S*-isomer of **18d**. Conversely, in the presence of **[H$_8$]-PA 12**, the *R*-isomer of **18d** was formed. Gong has also reported on a positive nonlinear effect with respect to this reaction among others, whereby racemic forms of catalysts are less soluble than enantioenriched versions leading to asymmetric amplification [84]. Comparative work by Lin and Wang [85] and Zheng and Zhang [86] has shown SPINOL-derived and

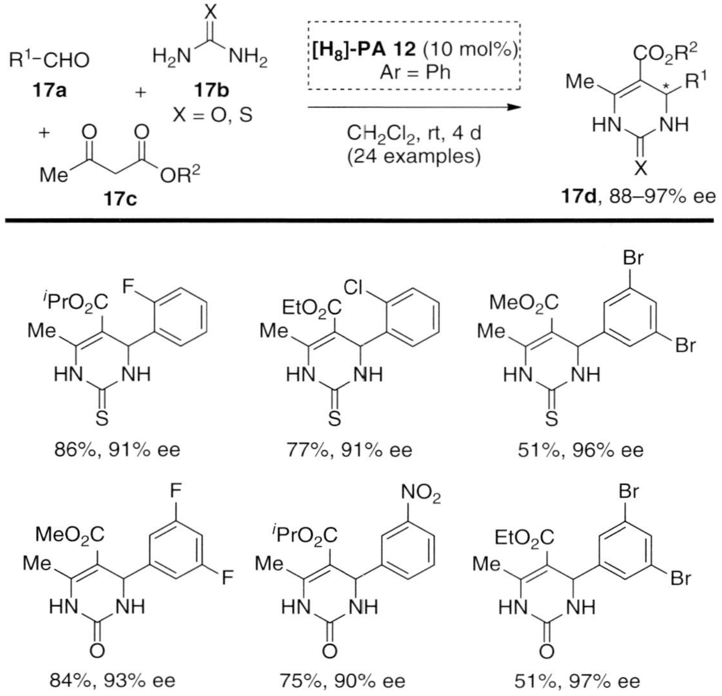

Scheme 2.35 Biginelli reaction by Gong [82].

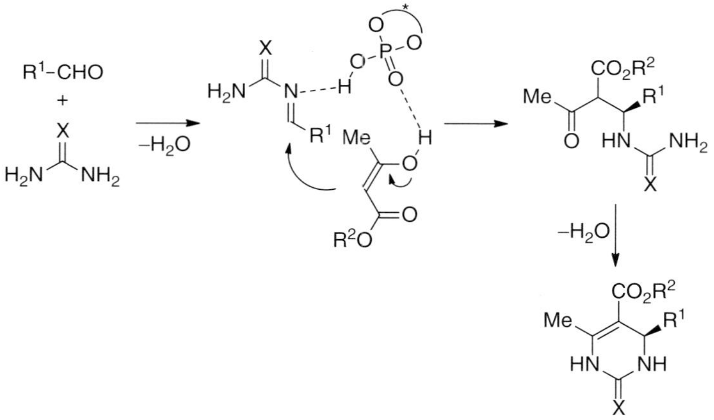

Scheme 2.36 Mechanism of the Biginelli reaction [82].

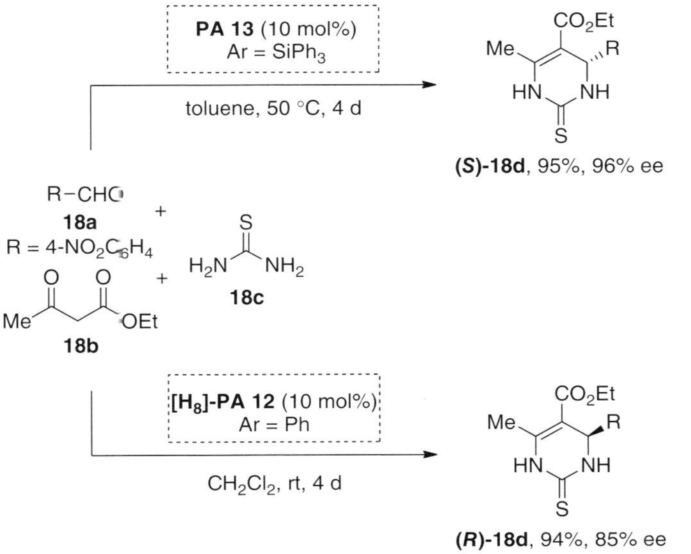

Scheme 2.37 Tuning absolute stereochemistry by catalyst choice by Gong [83].

bisphosphorylimide catalysts, respectively, to be suitable for also achieving high enantioselectivities.

2.5
Friedel–Crafts Reactions

2.5.1
Indole Coupling Partners with Aldimines

The area that Friedel–Crafts reactions accommodate within the field of asymmetric Brønsted acid catalysis is a considerable size. The success of the concept has encouraged many research groups to contribute to this area, and as such, a wealth of literature exists. The seminal report in this subsection came relatively late in 2007 from You, who reported on the coupling of tosyl imines **19a** with N–H-free indoles **19b** to give the corresponding products **19c** generally with excellent enantioselectivities (Scheme 2.38) [87, 88].

The reaction generally proceeds smoothly for aromatic imines, while when aliphatic imines were employed, lower enantioselectivities were observed. The mechanism follows the standard Friedel–Crafts process and can be observed in Scheme 2.39. Activation of both components occurs by the catalyst, which allows the indole to attack the imine. The C3 proton is then lost, and the product is subsequently formed.

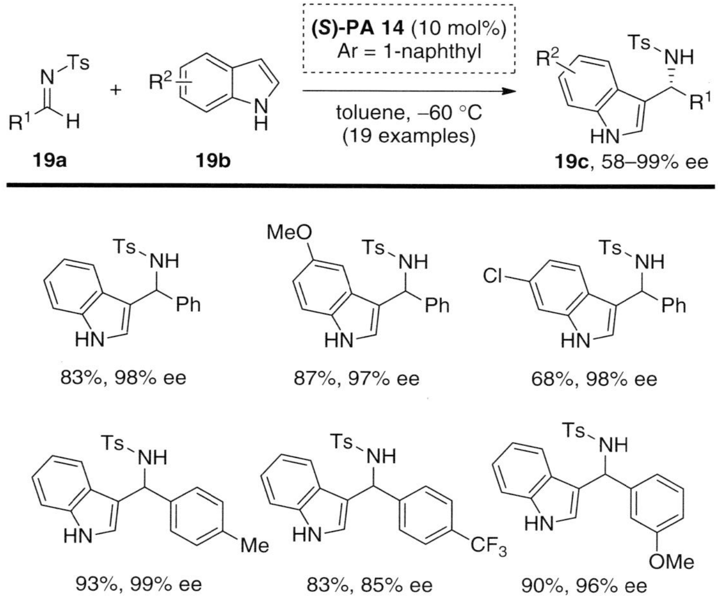

Scheme 2.38 Friedel–Crafts reaction of indoles with aldimines by You [87].

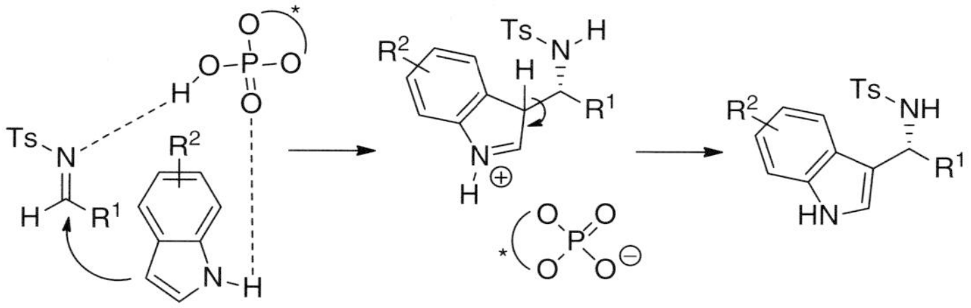

Scheme 2.39 Mechanism of the Friedel–Crafts reaction [87].

A range of other groups have also researched into this reaction and have utilized chiral phosphoric acids as suitable catalysts [89–95]. This reaction has also encouraged groups to develop new catalysts, and as such, a variety has been reported based on different scaffolds (Figure 2.2).

Lee has investigated into a chiral sulfonimide catalyst (**BA 9**), which generally provides good selectivities but is intolerant to alkyl imines [96]. In 2010, Terada developed a tartaric-acid-based catalyst (**BA 10**); however, only low to modest enantioselectivities could be achieved [97]. Finally, Jiang and Zhang have described a double axially chiral bisphosphorylimide catalyst (**NPA 2**), which shows extremely high activity for both aryl and alkyl imines [98].

A rather unique example of the utilization of a Friedel–Crafts reaction was shown by Enders in a two-step cascade process toward the synthesis of isoindolines. Suitably designed imine substrate **20b** can be reacted with indole **20a** in the

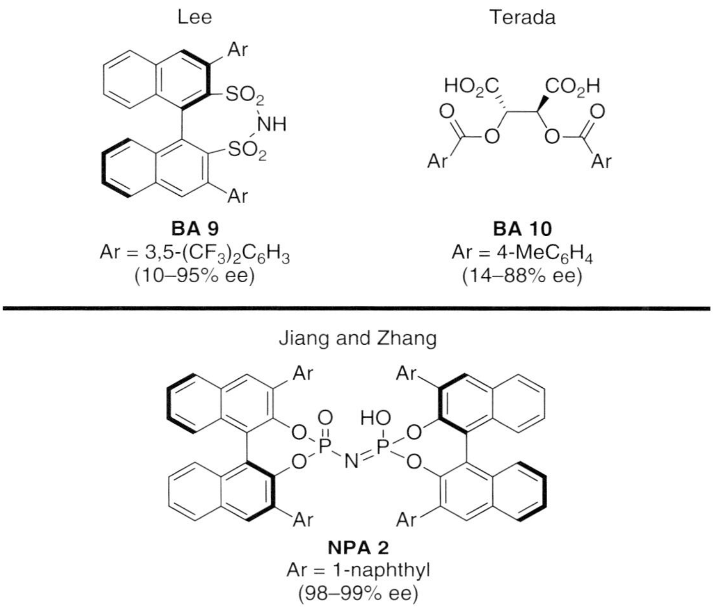

Figure 2.2 Catalysts used for the Friedel–Crafts reaction of aldimines.

presence of *N*-trifyl phosphoramide **NPA 3** to initially give the Friedel–Crafts adduct, which could be cyclized by treating with DBU to give the products **20c** with generally high enantioselectivities (Scheme 2.40) [99].

An enantioselectivity enhancement was noticed during the first step of the reaction if the reaction was allowed to run for longer periods and was proposed to be due to preferential overreaction of one enantiomer, thus lowering the yield of the desired product.

The mechanism of the cascade is depicted in Scheme 2.41. The first stage of the process follows the standard mechanism as shown in Scheme 2.39, whereby the imine and indole are both activated and 1,2-addition occurs. After loss of a proton, a benzylic amine intermediate is formed, which can go on to cyclize onto the neighboring alkene in a 1,4-addition manner to yield the isoindoline products. A closely related transformation although not formally a Friedel–Crafts reaction has also been disclosed by the group to access tetrahydroisoquinolines [100].

2.5.2
Indole Coupling Partners to Ketimines

Ketimine substrates pose a unique challenge for asymmetric Brønsted acid catalysis as more often than not they behave in very different modes. This usually means that the methodology optimized for aldimines does not always translate well to ketimine substrates. Goodman has highlighted this difference by careful examination of different imines toward undergoing Friedel–Crafts reactions [101].

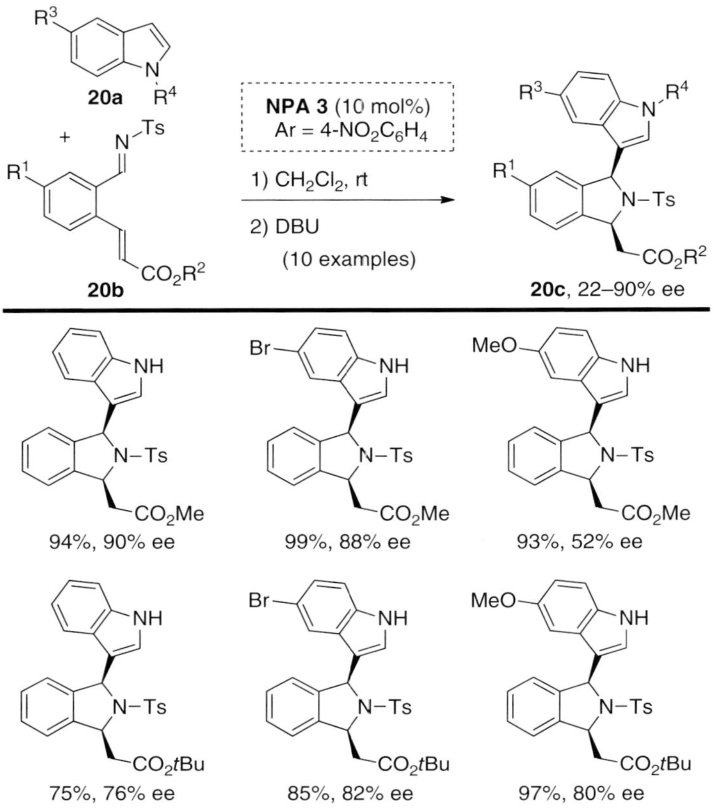

Scheme 2.40 Two-step Friedel–Crafts/Michael reaction by Enders [99].

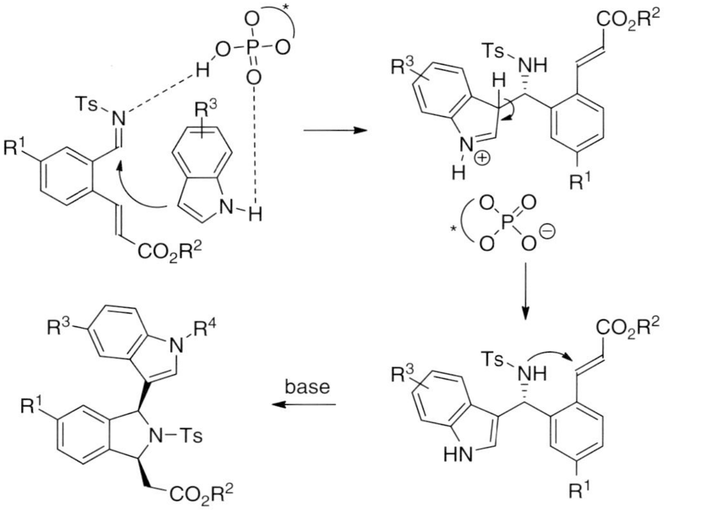

Scheme 2.41 Mechanism of the Friedel–Crafts cascade [99].

In 2011, Rueping was able to show that aromatic cyclic imine substrates **21b** could couple with indoles **21a** in the presence of **[H$_8$]-PA 13** to give chiral indolindolinones **21c** with good selectivities (Scheme 2.42) [102]. The use of cyclic aromatic imines has also been shown by other research groups [103–105].

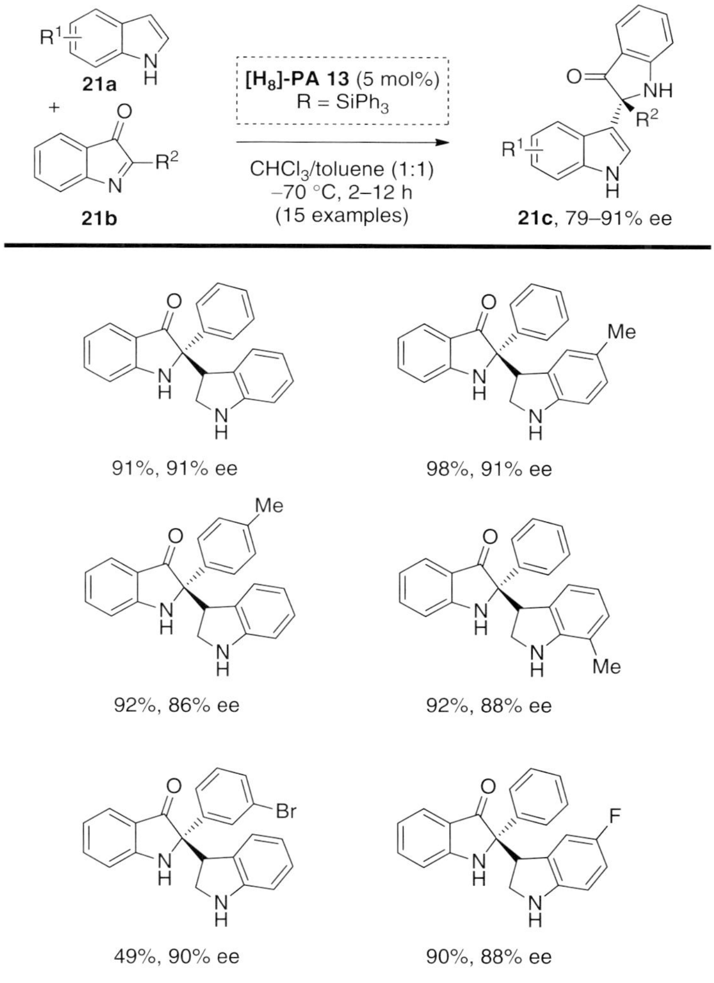

Scheme 2.42 Friedel–Crafts reaction of ketimines by Rueping [102].

In 2012, ketimines derived from isatins **21e** were shown by Wang to successfully react with various aromatic nucleophiles **21d** to yield the corresponding products **21f** with excellent enantioselectivities (Scheme 2.43) [106]. It was found that by the addition of 4 Å molecular sieves to capture residual water, the enantioselectivity was greatly improved.

The asymmetric construction of quaternary amino acid derivatives is a highly sought-after moiety, and by the appropriate use of ketimines under Brønsted acid

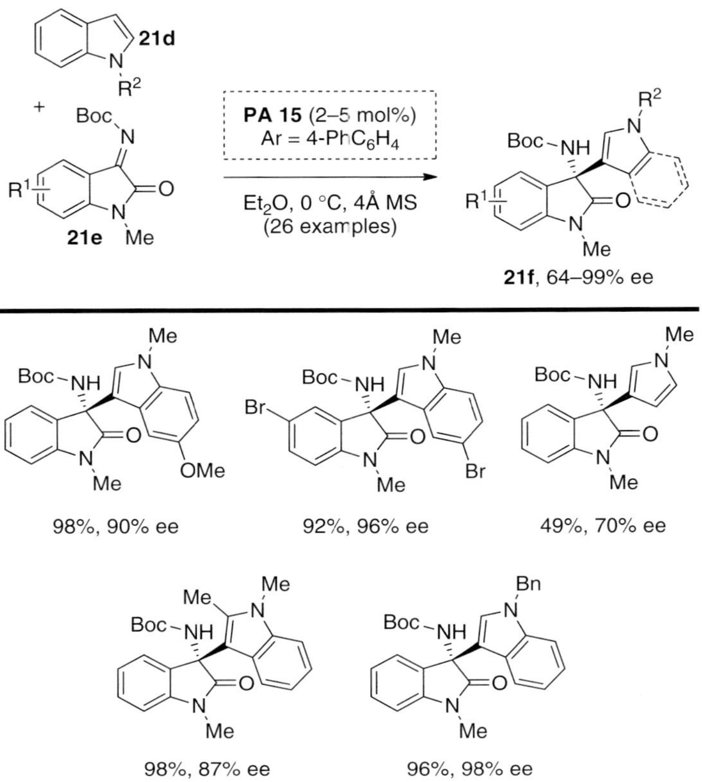

Scheme 2.43 Friedel–Crafts reaction of ketimines by Wang [106].

catalysis, this challenge has been overcome to some part. In 2011, Bolm was able to realize the use of fluorinated pyruvate imine **22b** with various indoles **22a** to yield the coupling products **22c** (Scheme 2.44) [107].

Interestingly, the reaction provides only low yields when less sterically hindered catalysts are used; however, the use of **PA 16** delivered high levels of enantioselectivity. Hu has also shown the use of pyruvate imines in a similar transformation to yield quaternary center containing amino acid derivatives [108].

2.5.3
Non-Indole Aromatic Partners

Historically, indoles are known to perform extremely well under Brønsted acid catalysis, and so, the use of non-indole partners is scarcely seen. One of the first reports concerning the Friedel–Crafts reaction was by Terada, who showed the use of 2-methoxyfuran **23b** as a suitable partner for Boc-protected imines **23a** to yield the products **23c** with high selectivities (Scheme 2.45) [109].

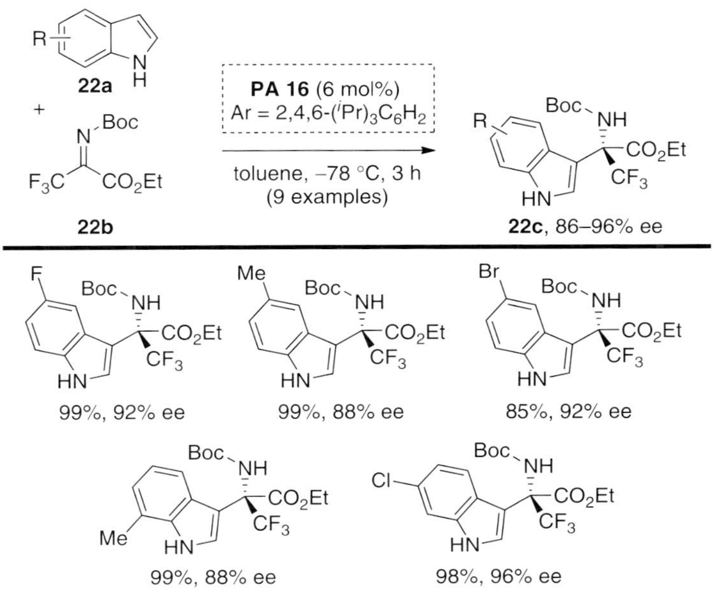

Scheme 2.44 Friedel–Crafts reaction of CF_3-containing ketimines by Bolm [107].

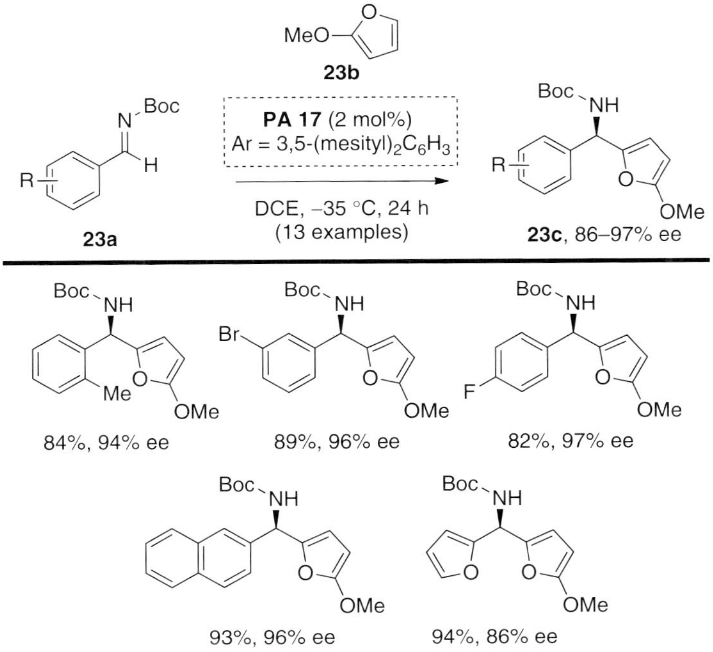

Scheme 2.45 Friedel–Crafts reaction of non-indole aromatics by Terada [109].

The benefit of using furan as a coupling partner is that it is a highly versatile compound that can be transformed into various functional groups, which was also shown by Terada. A few years later, Antilla published a comparable study on the use of substituted pyrroles **23e** with acyl imines **23d** to give the products **23f** with high enantioselectivities (Scheme 2.46) [110]. Noteworthy was the fact that N–H-free pyrroles yielded the desired products but with almost no enantioselectivity. This result suggests that coordination of the pyrrole partner must compete comparatively with a pathway involving no coordination. The groups of Nakamura [111], You [112], Ishihara [113], and Jiang and Zhang [114] have also shown the use of pyrroles and dihydroindoles in similar transformations.

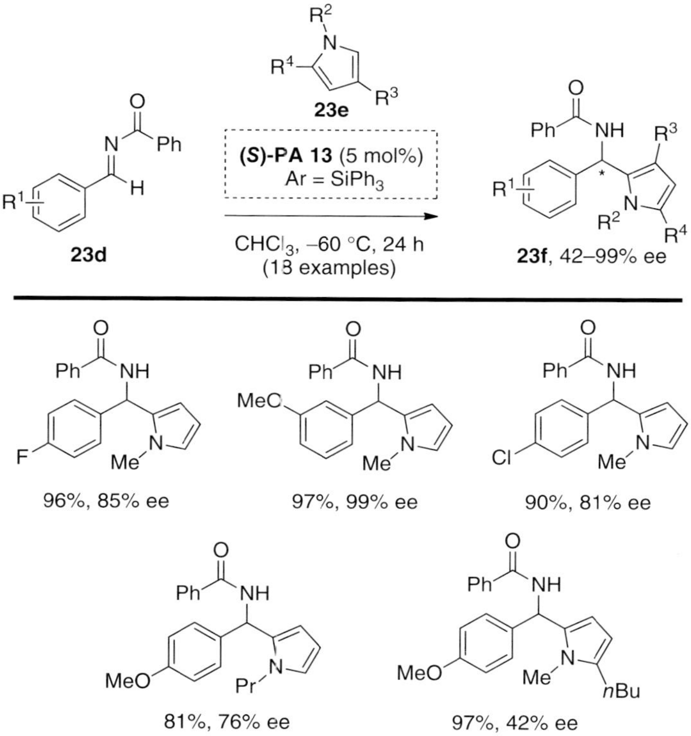

Scheme 2.46 Friedel–Crafts reaction of non-indole aromatics by Antilla [110].

A more classical Friedel–Crafts reaction was reported by Enders in 2010, which employed various benzene derivatives **24a** with glyoxylate imine **24b** to yield the products **24c** with good to modest selectivities (Scheme 2.47) [115].

Interesting to note was that most phosphoric acid catalysts that were tested provided poor enantioselectivities and even the corresponding thiophosphoric acid

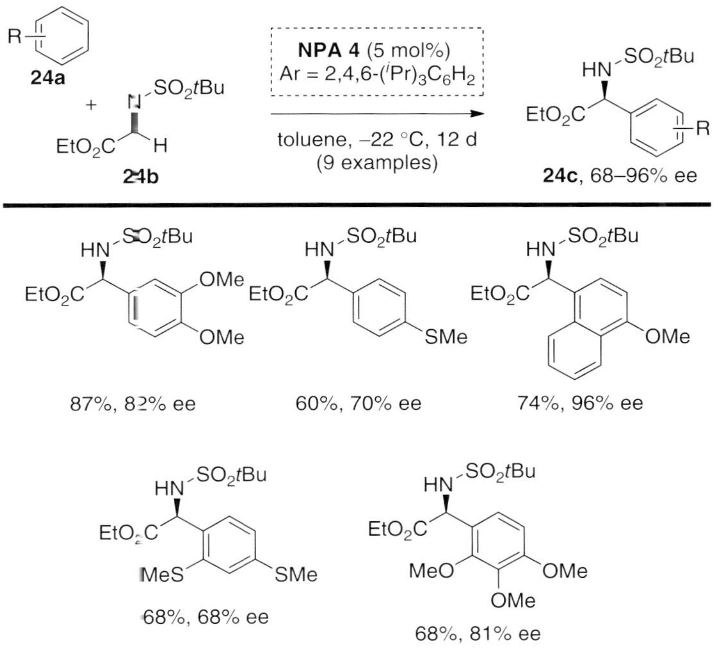

Scheme 2.47 Friedel–Crafts reaction of benzenes by Enders [115].

of **NPA 4** yielded poor results. The sulfonyl protecting group was shown by the group to be easily removed without any racemization, allowing access to amino acid derivatives.

2.5.4
Pictet–Spengler Reactions

The Pictet–Spengler reaction is a cascade process that represents an important subsection within Friedel–Crafts reactions for the synthesis of polycyclic motifs. The first development of an asymmetric process catalyzed by a Brønsted acid came from the List group in 2006, which showed that tryptamines **25a** could react with aldehydes **25b** to yield tetrahydro-β-carboline products **25c** with high enantioselectivities (Scheme 2.48) [116].

It was shown that the presence of a doubly substituted ester carbon center was a requirement, which aided the cyclization process to occur. The esters, however, could be diastereoselectively functionalized, allowing access to a further stereocenter with excellent selectivity. A simplified mechanism for this process can be thought of occurring as illustrated in Scheme 2.49. Condensation of the aldehyde with the tryptamine should lead to an imine intermediate, which can be activated by the catalyst to undergo addition at the C2-position of the indole

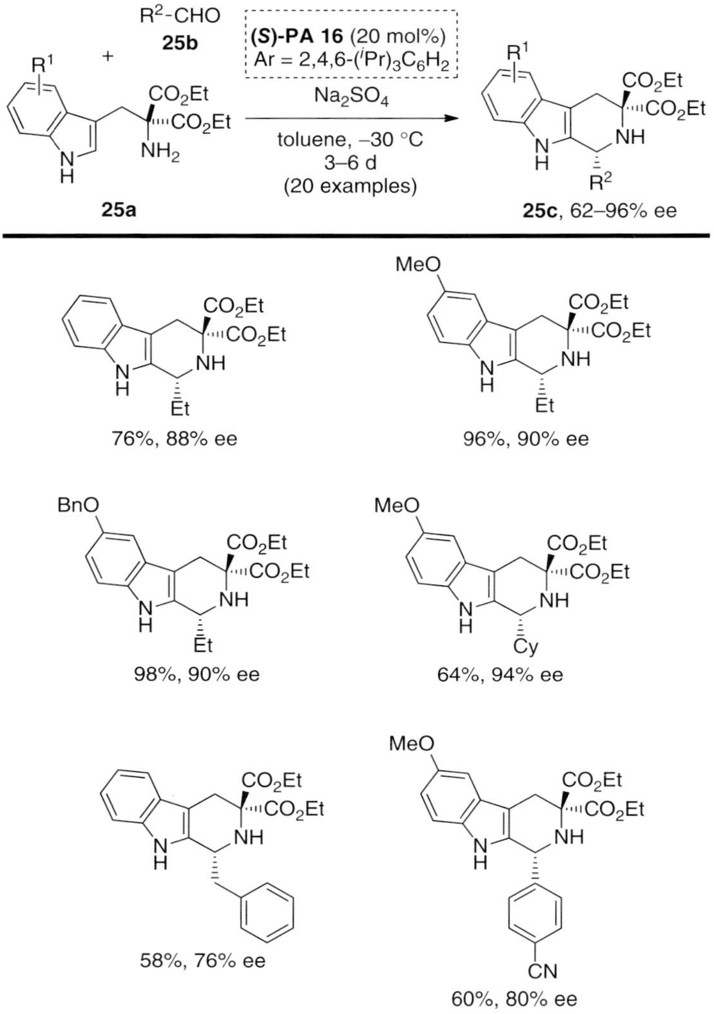

Scheme 2.48 Pictet–Spengler reaction by List [116].

to yield the desired product. An alternative mechanism can also be envisioned, which involves addition to the C3-position and then subsequent migration to the C2-position.

The necessity of a having a fully substituted center to aid the cyclization process has been eliminated by Hiemstra by the use of *N*-sulfenyltryptamine substrates [117]. The group has also used this reaction to synthesize a variety of natural

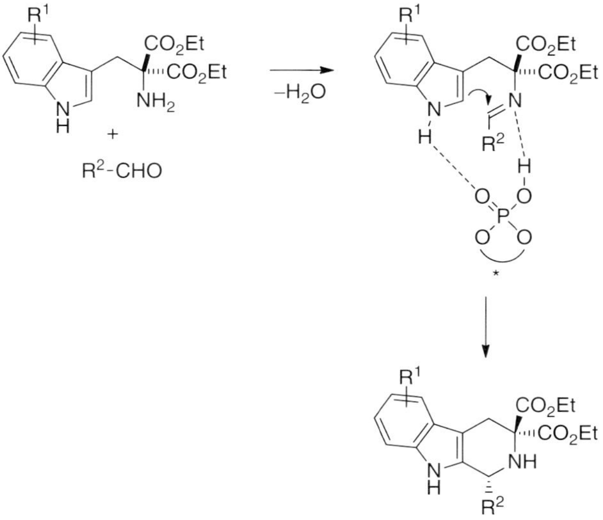

Scheme 2.49 Mechanism of the Pictet–Spengler [116].

products [118–121]. Closely related extensions to include different substrates and catalysts have also been developed [122–126]. An interesting extension of the classical Pictet–Spengler has been disclosed by Bernardi and Bencivenni, who employed isatins **26a** as coupling partners to tryptamines **26b** to access enantioenriched spiroindolinones **26c** with good levels of control (Scheme 2.50) [127].

Quite unusually, the solvent DMF was found to be the optimal medium and could even be used at reagent-grade quality. The enantioselectivity also remained intact in the presence of moisture, although the conversion did suffer. A related transformation in the same year was disclosed by Franz, who also employed DMF as the optimal solvent [128].

Chiral Brønsted acid catalysts are well known for the ability to control asymmetric reactions, but simple acid catalysts also play an important role in catalyzing a whole host of useful transformations. In 2009, Dixon reported an exquisite cascade process involving *N*-acyl iminium ions generated *in situ* from lactones **27a** and tryptamines **27b**. The formation and subsequent stereoselective transformation are both catalyzed by **PA 13**, which leads to highly complex products **27c** with good enantioselectivities (Scheme 2.51) [129].

The starting lactones **27a** could also be formed *in situ* with an initial gold catalyzed step before the addition of the catalyst and tryptamine components. The Dixon group has also explored various extensions to this methodology [130–132].

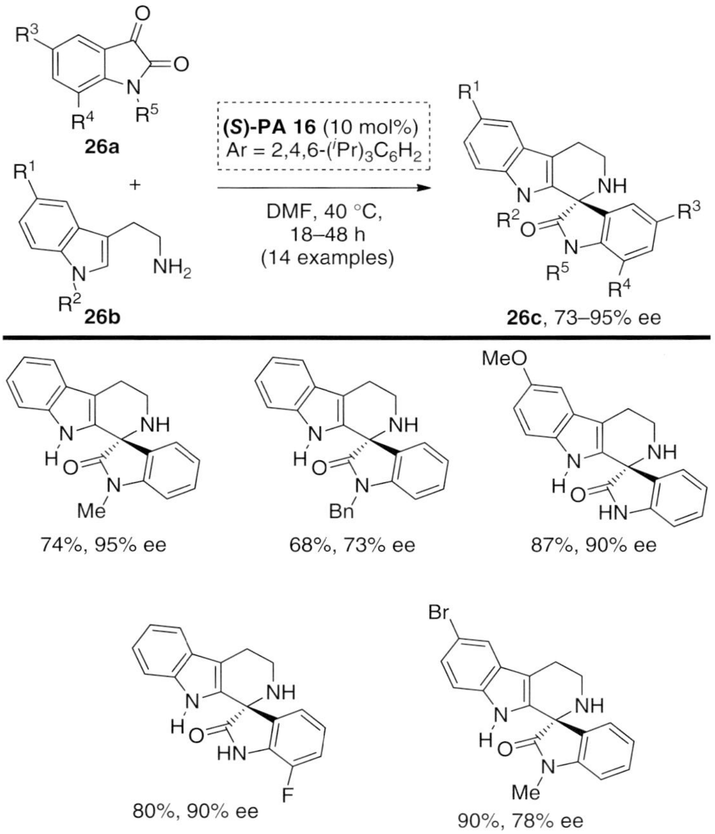

Scheme 2.50 Pictet–Spengler reaction to access spiroindolinones by Bernardi and Bencivenni [127].

2.6 Transfer Hydrogenations

2.6.1 Imine Partners

Transfer hydrogenations have received a great amount of attention with regard to protocols concerning chiral Brønsted acids. Even though the starting materials usually contain imine-like motifs, this has not reduced their appeal to research groups. The simple asymmetric reduction of imines represents one of the most straightforward methods to access chiral amine products. The groups of Rueping and List developed the first protocols for the asymmetric reduction of imines using chiral phosphoric acids and employed Hantzsch esters as a reducing

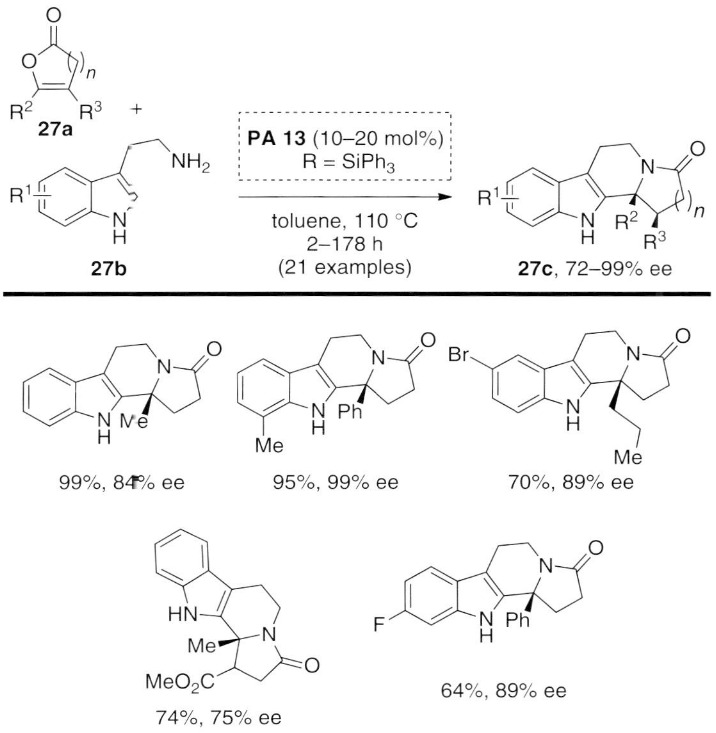

Scheme 2.51 Pictet–Spengler cascade reaction by Dixon [129].

source. Rueping showed that various aryl-protected imines **28a** could be reduced smoothly to amines **28b** in benzene at 60 °C (Scheme 2.52) [133].

List demonstrated that PMP-protected imines **28c** could undergo reductions to the corresponding amines **28d** in toluene at just 35 °C (Scheme 2.53) [134]. Although most imines utilized were aromatic, List also showed one example of an alkyl imine that performed comparatively well. Protected and unprotected imines have also been shown to be viable substrates under similar reaction conditions by Wang [135–137]. The mechanism of the reduction involving Hantzsch esters is described in Scheme 2.54. Activation of both components occurs by the catalyst, which leads to the selective transfer of hydride to the imine. The catalyst is then reformed by proton abstraction from the protonated pyridine intermediate. This general mechanism can be considered to be occurring irrespective of the hydride donor, of which several will be encountered in this section.

While imine substrates resemble incredibly versatile components for Brønsted acid catalysis, their inherent instability detracts somewhat from their usage. The reductive amination reaction is one procedure that allows the user to

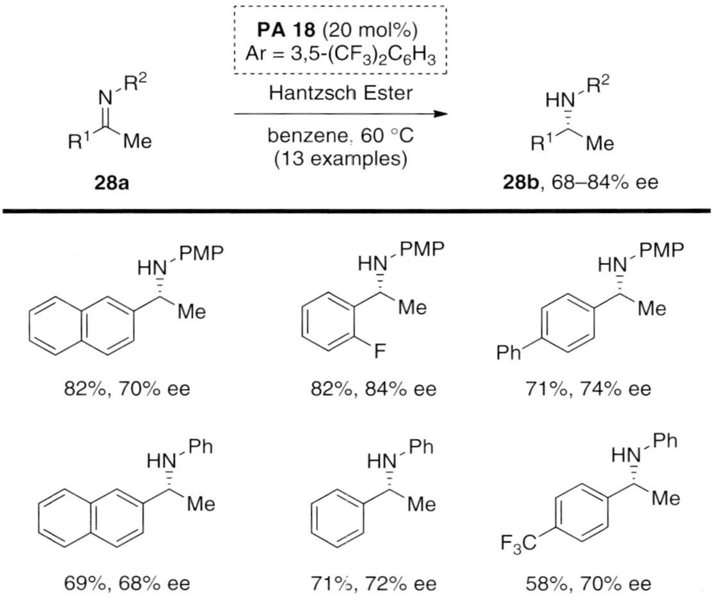

Scheme 2.52 Transfer hydrogenation of imines by Rueping [133].

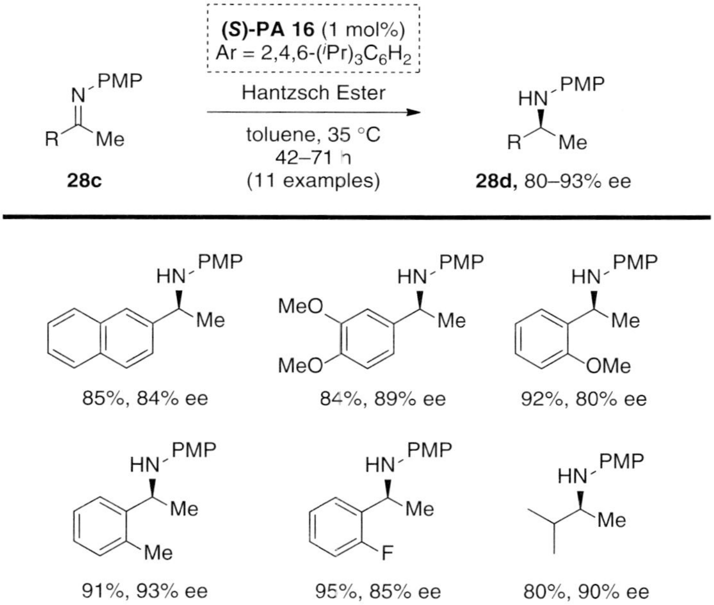

Scheme 2.53 Transfer hydrogenation of imines by List [134].

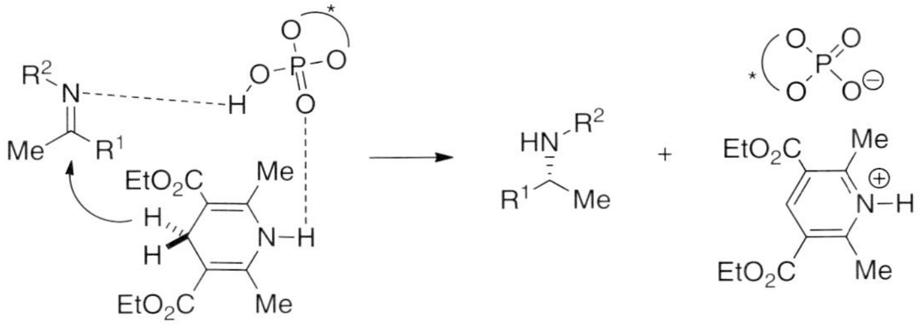

Scheme 2.54 General mechanism of transfer hydrogenation of imines.

circumvent this problem and achieve the same end products. To this extent, MacMillan has developed an enantioselective reductive amination using aromatic amine **29a** and aromatic ketones **29b** in the presence of a Hantzsch ester to yield the secondary amine products **29c** with excellent enantioselectivities (Scheme 2.55) [138].

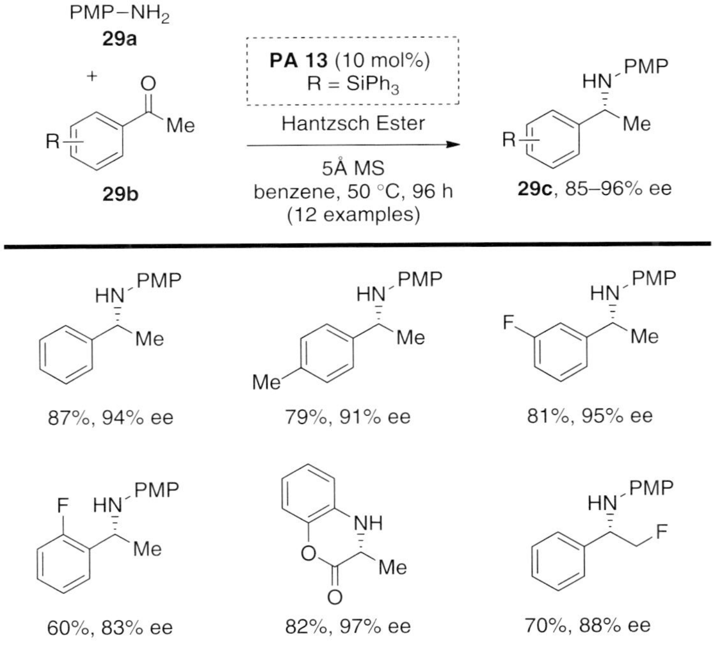

Scheme 2.55 Reductive amination by MacMillan [138].

The generated H_2O upon condensation needs to be captured, and so, the use of 5 Å molecular sieves was shown to perform this duty efficiently. The reaction proceeds smoothly for a range of substrates and also works for one alkyl substrate albeit with a lower level of enantioselectivity. List has been highly prolific in the development of asymmetric reductive amination protocol and a series of extensions have also been published by the group [139–141].

Although Hantzsch esters feature in the majority of reports concerning asymmetric transfer hydrogenations, several groups have shown other sources to be comparable in achieving good conversions and selectivities. Akiyama has been the major developer of benzothiazoline-reducing agents. In 2009, he reported the first protocol for the reduction of aryl-protected imines **30a** by benzothiazoline **30b** in the presence of **PA 16** to give the products **30c** with excellent enantioselectivities (Scheme 2.56) [142].

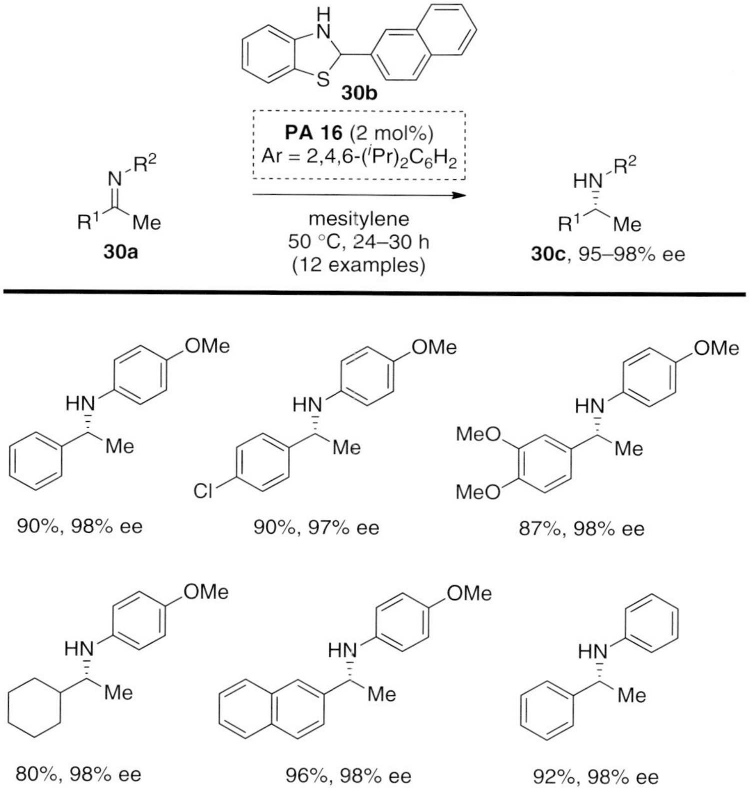

Scheme 2.56 Transfer hydrogenation using a benzothiazoline by Akiyama [142].

A broad scope that included an alkyl example was reported, and in addition, the benzothiazoline could also be generated *in situ*. Akiyama has extended his methodology to include fluorinated substrates [143, 144], a protocol for enantioselective deuteration, [145] and a focus on alkyl imines [146, 147]. A closely related procedure has also been used by Shimizu for the synthesis of chiral β-lactams [148].

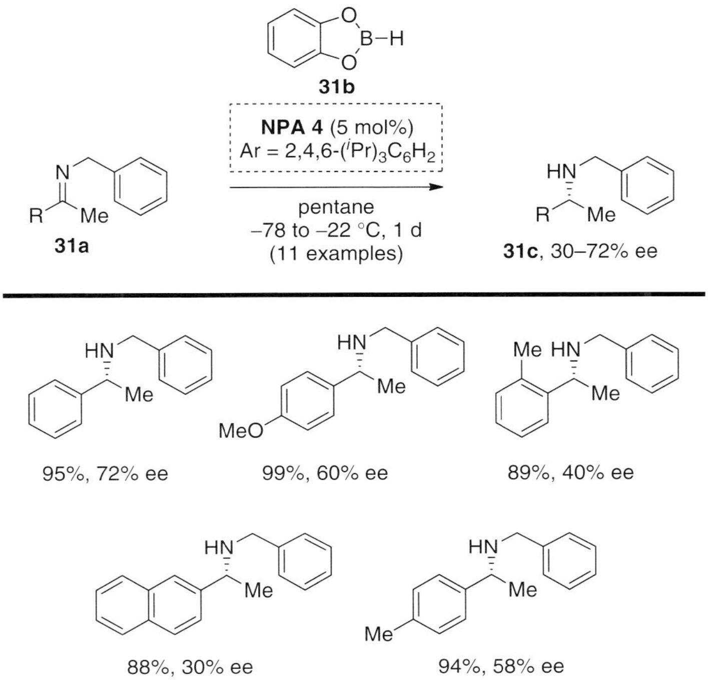

Scheme 2.57 Transfer hydrogenation using catecholborane by Enders [149].

In 2013, Enders has shown that catecholborane **31b** could be utilized as a reducing agent for benzyl-protected imines **31a** to yield the products **31c** with modest levels of selectivity (Scheme 2.57) [149]. In this case, phosphoric acid catalysts failed to deliver acceptable levels of enantioselectivity and so **NPA 4** was found to be suitable.

Lastly, Akiyama has recently found that indolines can serve as reducing agents for imine substrates. Taking aryl-protected imines **32a** with indoline **32b** in the presence of **PA 16** gave the corresponding reduced products **32c** with exceptional enantioselectivities (Scheme 2.58) [151, 150].

The process was developed for aromatic imines and also the reductive amination of aliphatic substrates. Furthermore, the indole by-product formed was shown to be easily recycled via a simple reduction procedure.

2.6.2
α-Imino Ester Partners

α-Imino ester substrates form an important subsection of imines since their reduction leads directly to amino acid derivatives, which are important to

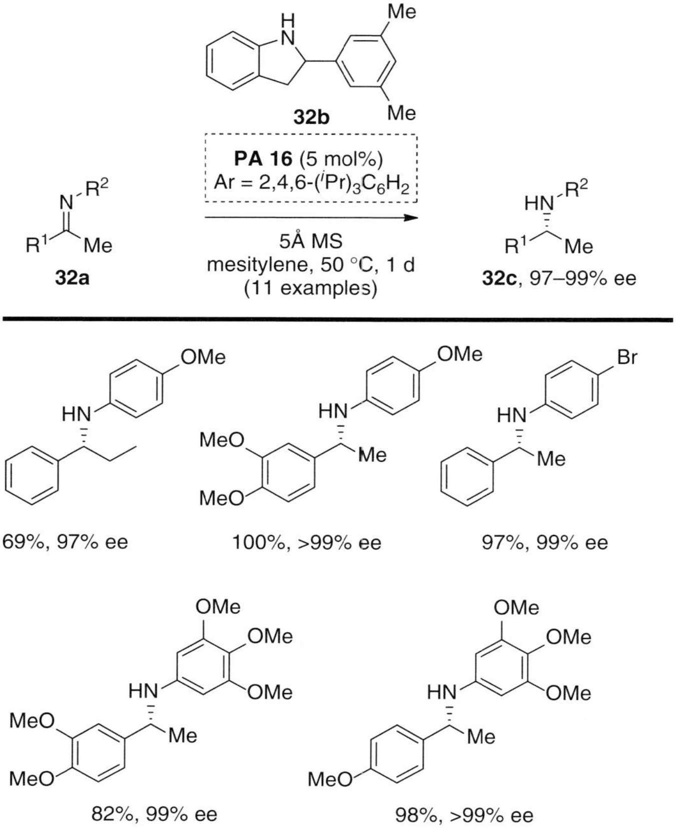

Scheme 2.58 Transfer hydrogenation using an indoline by Akiyama [150].

many industries. The first general procedure of this transformation was reported by Antilla, who showed that α-imino esters **33a** could be reduced by Hantzsch ester in the presence of VAPOL-derived catalyst **(*S*)-PA 6** (Scheme 2.59) [152].

The process functioned smoothly for a range of aromatic and alkyl imines and additionally *in situ* formation was also tolerated well for alkyl substrates in particular. A similar protocol has also been developed by You, which also includes β,γ-alkynyl α-imino esters [153, 154].

The reduction of α-imino esters has also been shown to be possible with alternatives to Hantzsch esters. In 2010, Akiyama demonstrated that benzothiazoline **34b** could be used to deliver **34c** with consistently excellent enantioselectivities (Scheme 2.60) [155]. Rather remarkably, only 1 mol% catalyst was required to achieve these high levels of selectivity.

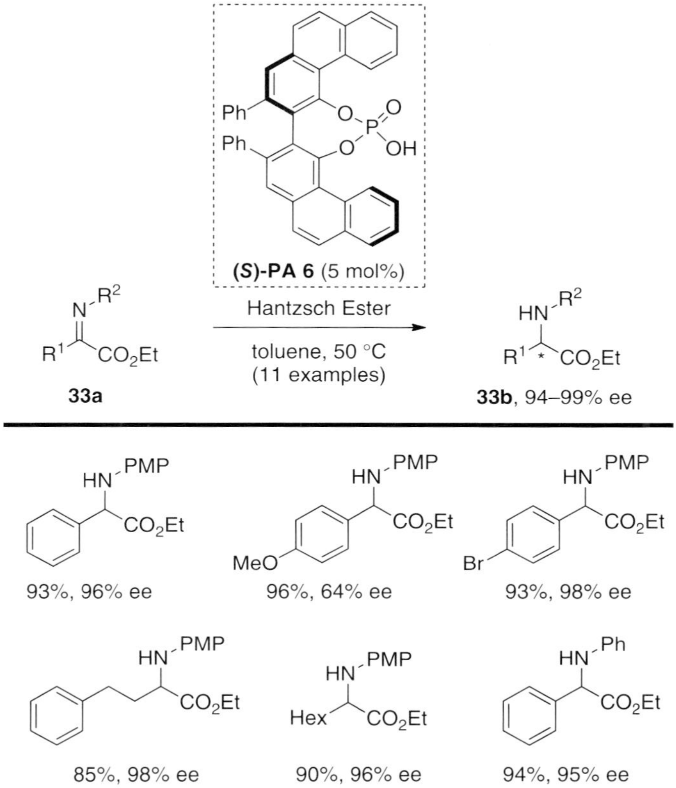

Scheme 2.59 Transfer hydrogenation of α-imino esters by Antilla [152].

Conversely, Enders has shown that catecholborane **34e** can be used to provide **34f** with modest enantioselectivities (Scheme 2.61) [156]. The protecting group of the imine was shown to have a significant influence on the selectivity, with a PMP group being the optimal solution.

2.6.3 *N*-Heterocycles

The asymmetric transfer hydrogenation of *N*-heterocycles is an area that has received a tremendous amount of attention. Analogous to imine substrates, the reduction of *N*-heterocycles represents a straightforward pathway to access these useful motifs, which are commonly found in bioactive molecules. The seminal study in this area came from Rueping, who found that quinolines **35a** could be effectively reduced with a Hantzsch ester in combination with catalyst **PA 4** (Scheme 2.62) [157]. The reduced products **35b** were obtained with high levels of enantioselectivity.

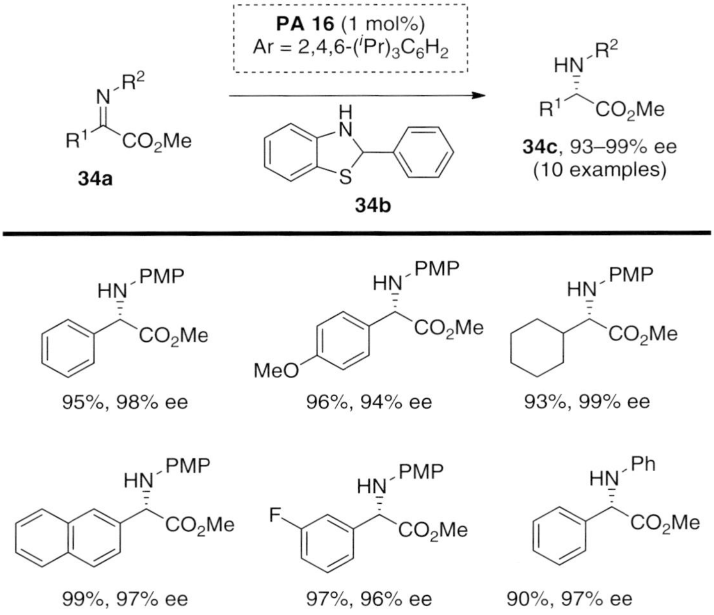

Scheme 2.60 Transfer hydrogenation of α-imino esters by Akiyama [155].

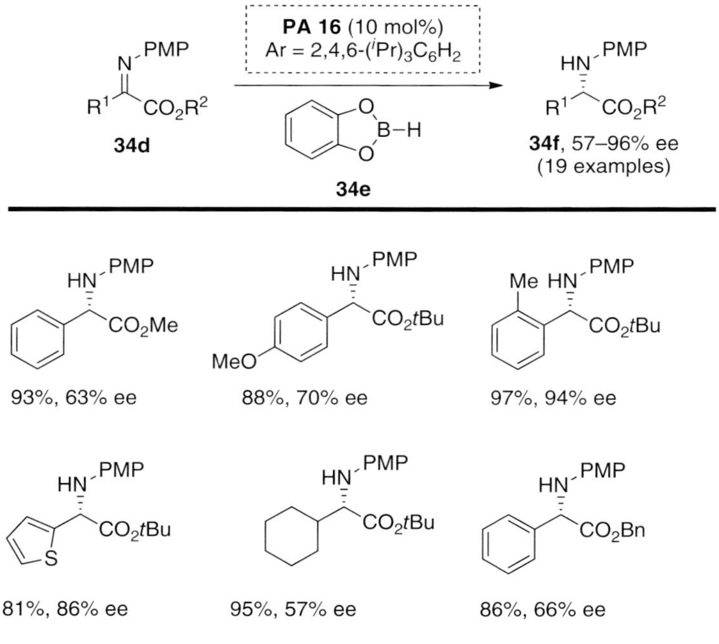

Scheme 2.61 Transfer hydrogenation of α-imino esters by Enders [156].

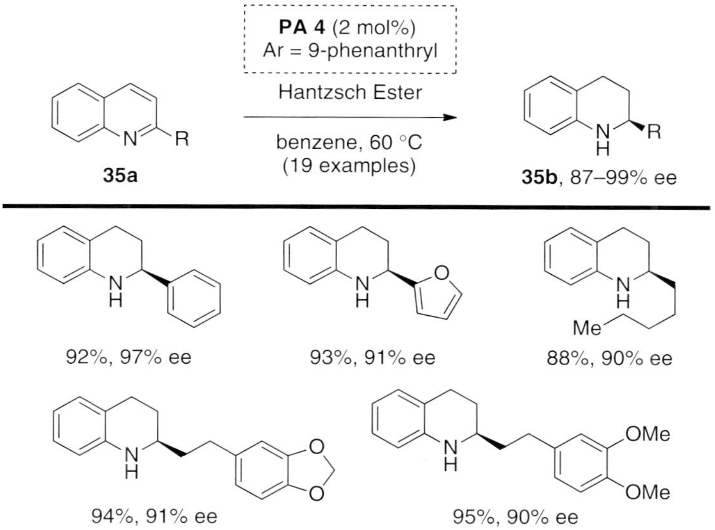

Scheme 2.62 Transfer hydrogenation of 2-substituted quinolines by Rueping [157].

The process works well for both aromatic and alkyl-substituted quinolines, and a short synthesis of three biologically active alkaloids was demonstrated. Since this report, the Rueping group has published extensions to this protocol [158, 159] including performing the procedure in H_2O [160] and in continuous flow [161]. Other research groups have also performed this reduction in the testing of new catalysts [162, 163], in particular polymer-supported catalysts [164–167]. The groups of Betzer and Marinetti have also developed catalysts based on unique architecture for performing the catalysis of quinoline reductions (Figure 2.3).

The earlier designs focus on planar [3,3]-paracyclophanes (**PA 19**), which are synthesized via straightforward steps from iodoferrocene [168, 169]. While recent

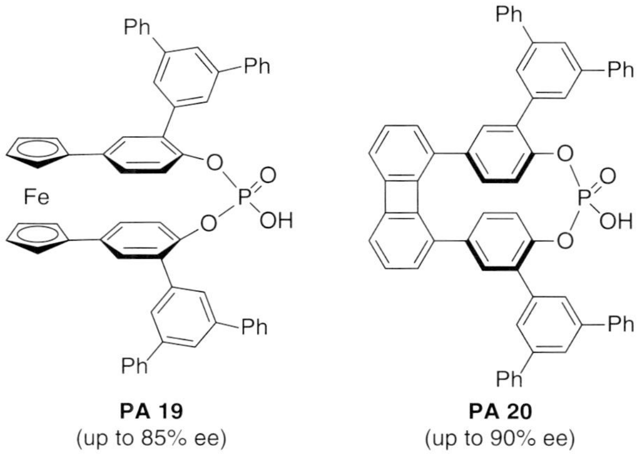

Figure 2.3 New catalysts for quinoline reduction by Betzer and Marinetti [168–170].

designs chose to incorporate a biphenylene unit into the catalyst design (**PA 20**) [170]. The asymmetric reduction of 2,3-disubstituted quinolines has also been shown [171–173].

Expansive research on quinoline reductions has seen groups investigate structurally similar heterocycles to explore the scope of asymmetric Brønsted acid catalysis. To that order, Metallinos has shown that 1,10-phenanthrolines can be reduced to yield **36a** with modest selectivities (Scheme 2.63) [174]. Rueping has

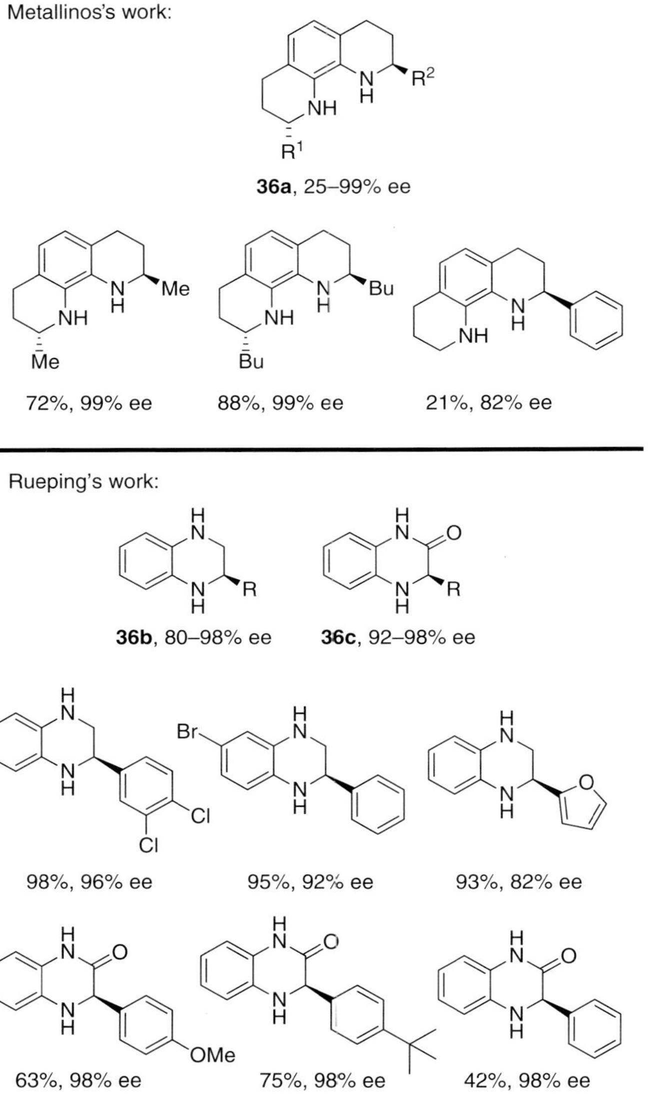

Scheme 2.63 Transfer hydrogenation of various *N*-heterocycles by Metallinos [174] and Rueping [175].

also shown that quinoxalines and quinoxalinones can be reduced efficiently to yield **36b** and **36c**, respectively, with high levels of selectivity [175]. Comparably, Shi and Tu have shown that asymmetric transfer hydrogenation can be performed on *in situ* formed quinoxalines from simple starting materials [176].

A further extension was also demonstrated by the Rueping group, which involved the reduction of benzoxazines and benzothiazines (**37a**) (Scheme 2.64) [177]. Using a Hantzsch ester, the reduction proceeded smoothly to yield the products **37b** with excellent enantioselectivities.

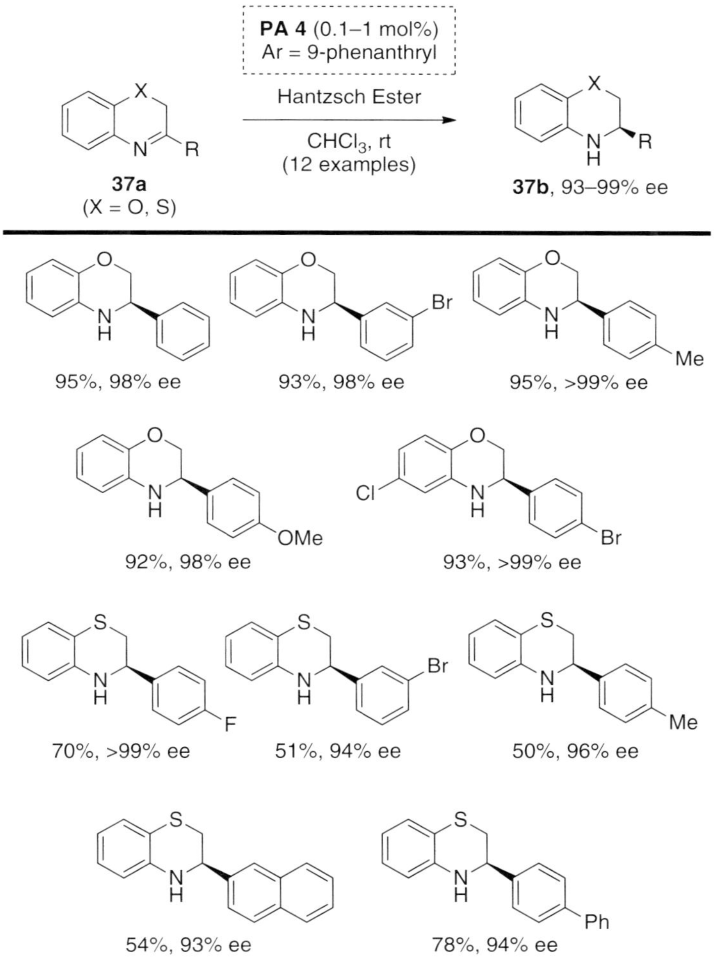

Scheme 2.64 The use of low catalyst loadings for transfer hydrogenations by Rueping [177].

Most remarkable about the procedure is that the catalyst loading used was 0.1 mol% in most cases but could even be dropped to 0.01 mol% without any significant drop in enantioselectivity. That level of catalytic activity is among the lowest reported within the field. In a related procedure, the reduction of fluoroquinolones has also been reported [178]. In 2007, the Rueping group disclosed important progress toward the enantioselective reductions of pyridines, which are notoriously difficult to achieve. Taking substrates **38a** with Hantzsch ester and catalyst **PA 1**, an enantioselective reduction took place to yield the tetrahydropyridine substrates **38b** with high levels of selectivity (Scheme 2.65) [179].

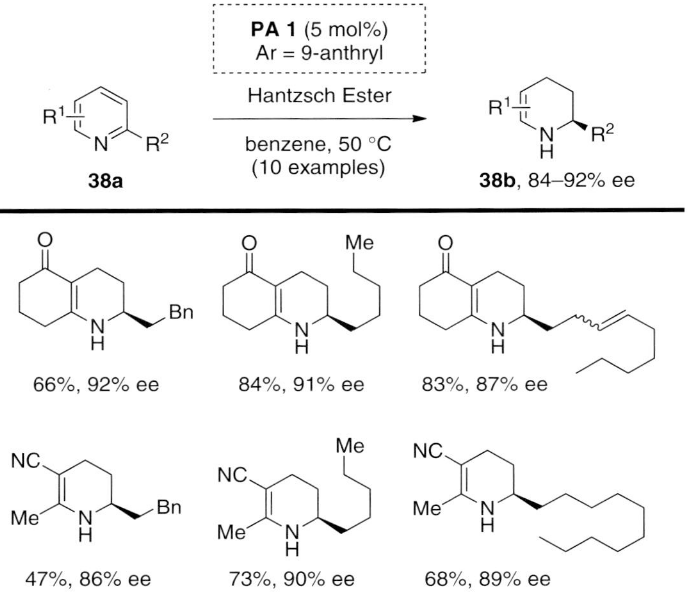

Scheme 2.65 Transfer hydrogenation of 2-substituted pyridines by Rueping [179].

It was shown that a carbonyl or a nitrile group was necessary at the 3-position in order to achieve suitable conversions and selectivities. Other heterocycles that have also featured in asymmetric transfer hydrogenation facilitated by Brønsted acids include diazepines, indoles, and benzodiazepinones (Scheme 2.66).

The Gong group has developed a dynamic kinetic transfer hydrogenation that leads to **39a** with modest selectivities [180], while the Rueping group has explored transfer hydrogenations to access indolines **39b** [181] and dihydro-benzodiazepinones **39c** [182] with good levels of enantioselectivities.

2.6.4
Cascade Processes

Since protons can catalyze a plethora of transformations, the development of cascade processes using Brønsted acid catalysts is generally an option open for

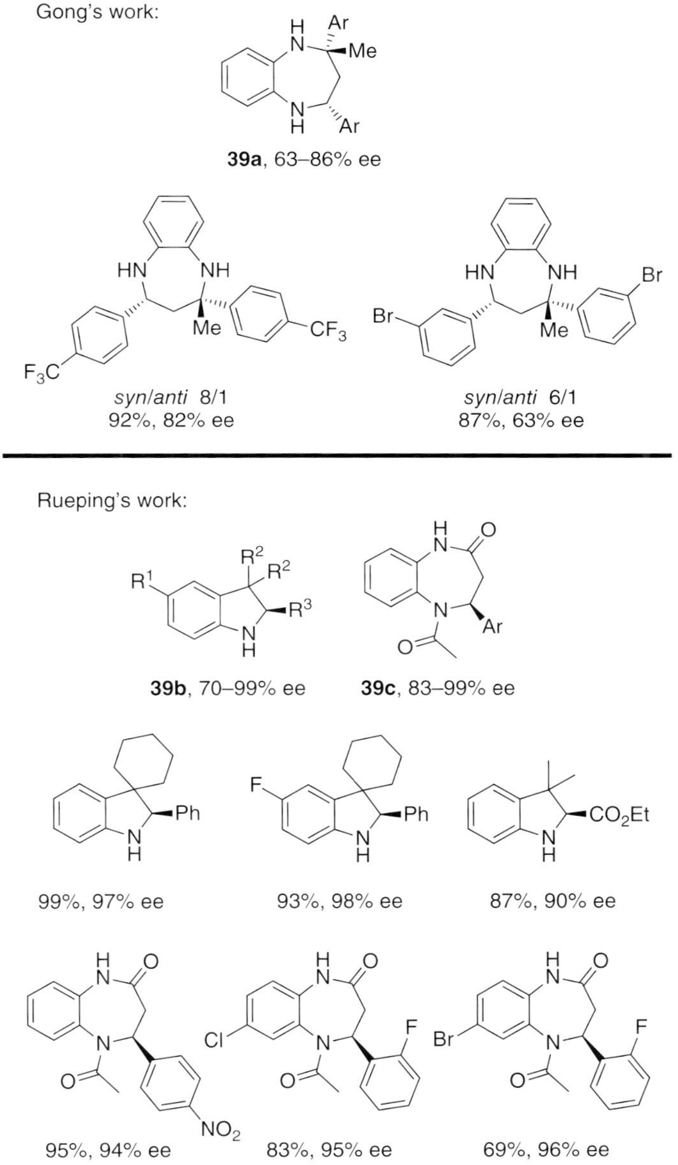

Scheme 2.66 Transfer hydrogenation of various *N*-heterocycles by Gong [180] and Rueping [181–182].

development; however, the challenge arises in controlling these transformations. In 2007, List realized the potential of using 1,5-dicarbonyls **40a** with an aromatic amine **40b** to undergo a controlled series of known transformations, which result in a final asymmetric transfer hydrogenation to arrive at 3-substituted cyclohexylamines **40c** (Scheme 2.67) [183].

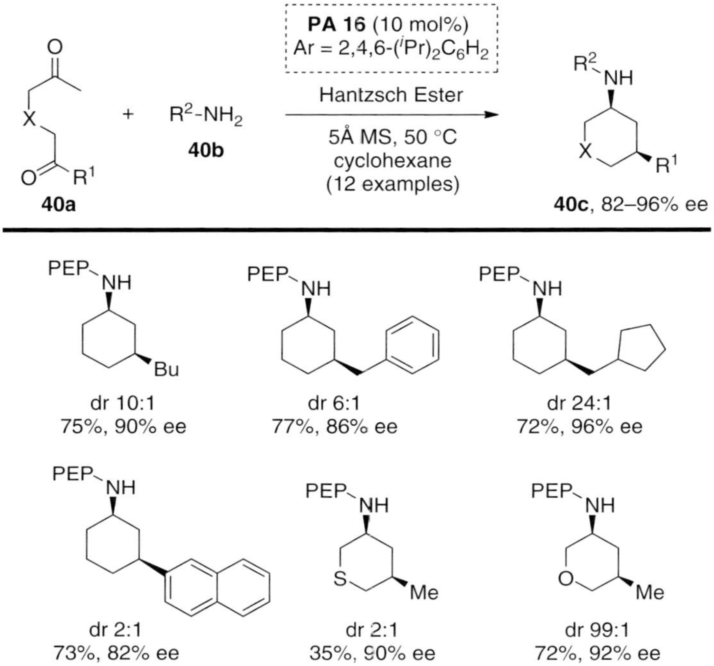

Scheme 2.67 Cascade transfer hydrogenation process by List [183].

The reaction provides the products **40c** with good diastereoselectivities and generally excellent enantioselectivities. Rather interestingly, **PA 16** gave the opposite diastereoisomeric products to a variety of other phosphoric acids used during the screening process. The mechanism of the process is given in Scheme 2.68. The cascade initially involves a series of condensations, firstly to form an enamine, which can go on to condense internally with the pendant ketone to yield an α,β-unsaturated imine. The condensations may also benefit from the chiral acid catalyst present even though no stereocenters are being formed. The catalyst's main role is to activate the α,β-unsaturated imine for a double reduction; firstly in a 1,4-addition of hydride and then a 1,2-addition to ultimately yield the products.

In 2011, Gong demonstrated a cascade process that involved a key asymmetric intramolecular transfer hydrogenation reaction. Condensation of ketones **41a** with aromatic amines **41b** led to an iminium intermediate, which would under

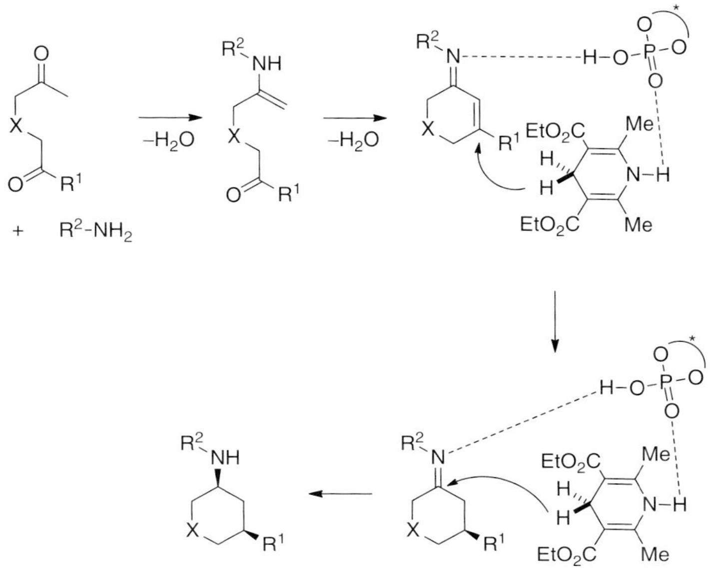

Scheme 2.68 Mechanism for cascade transfer hydrogenation process [183].

the influence of **PA 21** undergo an asymmetric 1,5-hydride shift followed by concomitant cyclization to yield **41c** with modest levels of selectivity (Scheme 2.69) [184].

A small sample of monophosphoric acid catalysts were evaluated for this process but failed to yield acceptable levels of selectivity. Two bisphosphoric acid catalysts were tested, which included an ether (**PA 21**) and thio-ether linkage; however, surprisingly, the thio-ether linkage gave comparatively poor enantioselectivity. The mechanism for the process can be observed in Scheme 2.70. The first step involves a condensation process, which reveals an imine and, once activated by the catalyst, allows for the intramolecular 1,5-hydride transfer step to occur. This generates a new iminium ion, which can be intercepted by the newly formed secondary amine to yield the desired products.

In 2013, the Rueping group showcased a rather novel combination of photochemistry and asymmetric Brønsted acid catalysis in the same reaction pot. They discovered that aminochalcones **42a** could undergo a light-assisted cyclative condensation to yield the representative quinoline substrates, which would undergo a transfer hydrogenation to yield the products **42b** with high enantioselectivities (Scheme 2.71) [185].

The mechanism for the cascade process is depicted in Scheme 2.72. The shining of light in the presence of the starting chalcone and the phosphoric acid is thought to cause an internal cyclative condensation, which leads to a cyclic iminium ion. The iminium ion also contains the bound phosphate, which helps to control the selectivity of the subsequent hydride transfer from Hantzsch ester.

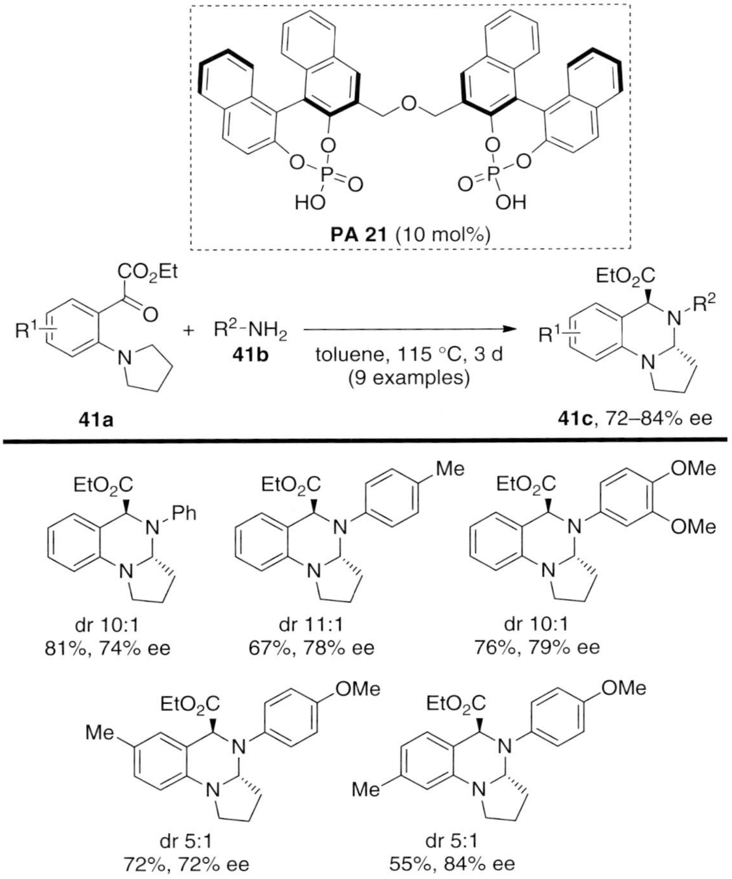

Scheme 2.69 Cascade hydride transfer-cyclization by Gong [184].

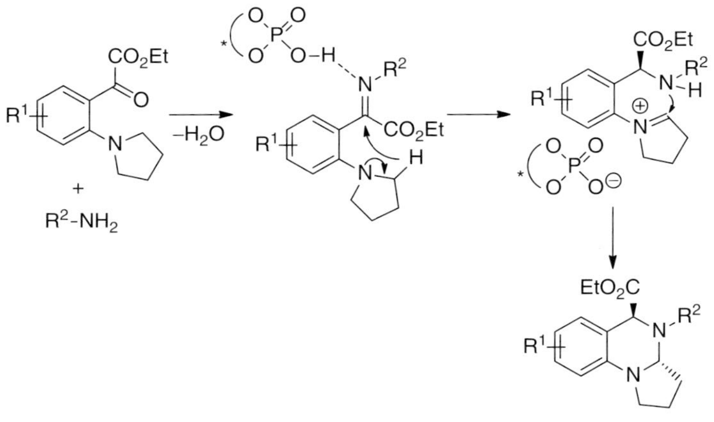

Scheme 2.70 Mechanism for the cascade hydride transfer cyclization [184].

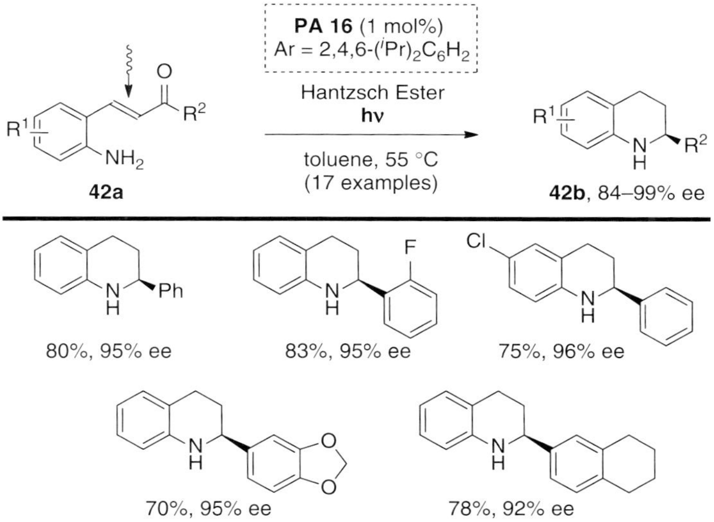

Scheme 2.71 A photocyclization–reduction cascade by Rueping [185].

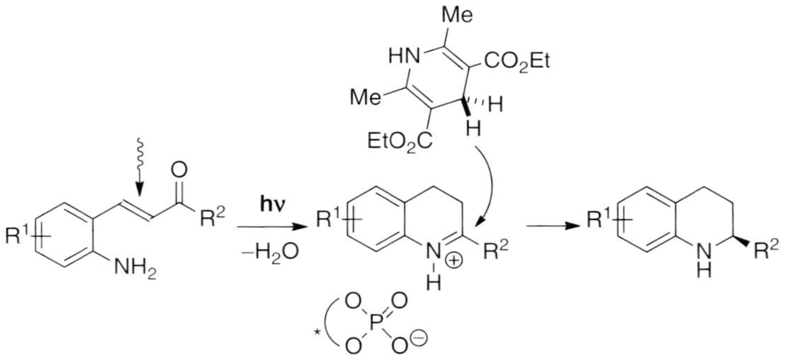

Scheme 2.72 Mechanism for photocyclization–reduction cascade [185].

The procedure developed tolerates a broad scope, and the unique harnessing of light with asymmetric catalysis should prove fruitful for future developments. One such advance was also shown by the use of a glass microreactor, which was able to absorb light more efficiently and thus led to a dramatic rate increase. The process was later shown to be amenable to be performed under continuous flow conditions [186].

2.6.5
Miscellaneous Reactions

In this section, we will cover the more unique examples of transfer hydrogenations in combination with Brønsted acid catalysts. In nearly all cases so far, the transfer hydrogenation of quinolines generates a stereocenter, which can be controlled by the catalyst. However, in 2008, Rueping recognized that positioning of a substituent at the 3-position of quinolines would lead to a stereocenter formed formally by protonation (Scheme 2.73) [187].

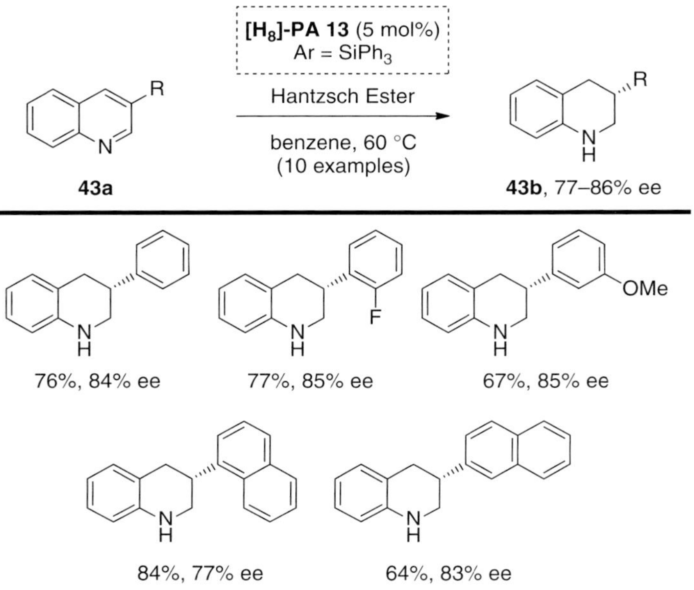

Scheme 2.73 Asymmetric protonation triggered by transfer hydrogenation by Rueping [187].

Taking substrates **43a** with catalyst **[H$_8$]-PA 13** resulted in a transfer hydrogenation to initially form an enamine intermediate, which could be asymmetrically protonated by the catalyst to deliver **43b** in modest levels of enantioselectivity. The Zhou group has also exploited this positioning effect to access 2,3-substituted quinolines [188, 189].

Whereas BINOL derivatives have found many applications in asymmetric catalysis, the corresponding BINAM derivatives have not featured as much, possibly due to the cost and availability of them in an enantiomerically pure format. Recently, Tian, Liu, and Tan have disclosed a unique route to access these motifs via a kinetic resolution reaction involving a transfer hydrogenation step (Scheme 2.74) [190].

Treatment of ***rac*-44** with an aldehyde, resulted in imine formation, and it was found that catalyst **PA 22** would preferentially activate the (*S*)-isomer in undergoing reduction by a Hantzsch ester, leaving the (*R*)-isomer **(*R*)-44** untouched with high levels of enantioselectivity. Removal of the protecting group was shown to be straightforward, yielding access to BINAM derivatives.

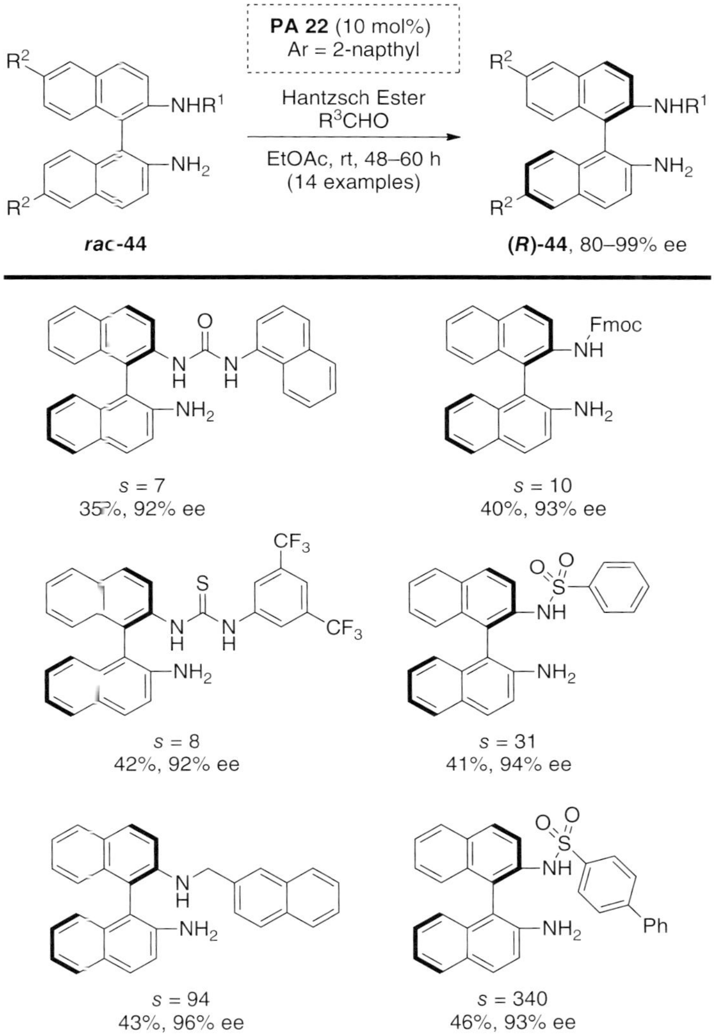

Scheme 2.74 Kinetic resolution of BINAM derivatives by Tian, Liu, and Tan [190].

2.7 Pericyclic Reactions

2.7.1 Aza-Diels–Alder/Povarov

The combination of two named reactions in this subsection has been chosen due to the similarities between the products obtained. Mechanistically, however, they formally follow very different pathways; namely a concerted versus a stepwise process. The protocols in this subsection have been categorized as Diels–Alder reactions, but one should note that this has only been used for namesake and does not infer any mechanistic assumptions.

In 2006, Akiyama reported a powerful two-step protocol for performing aza-Diels–Alder reactions with aryl-protected imines **45a** and Brassard's diene **45b** (Scheme 2.75) [191, 192]. The first addition step is performed at −40 °C, followed by the use of benzoic acid to complete the cyclization.

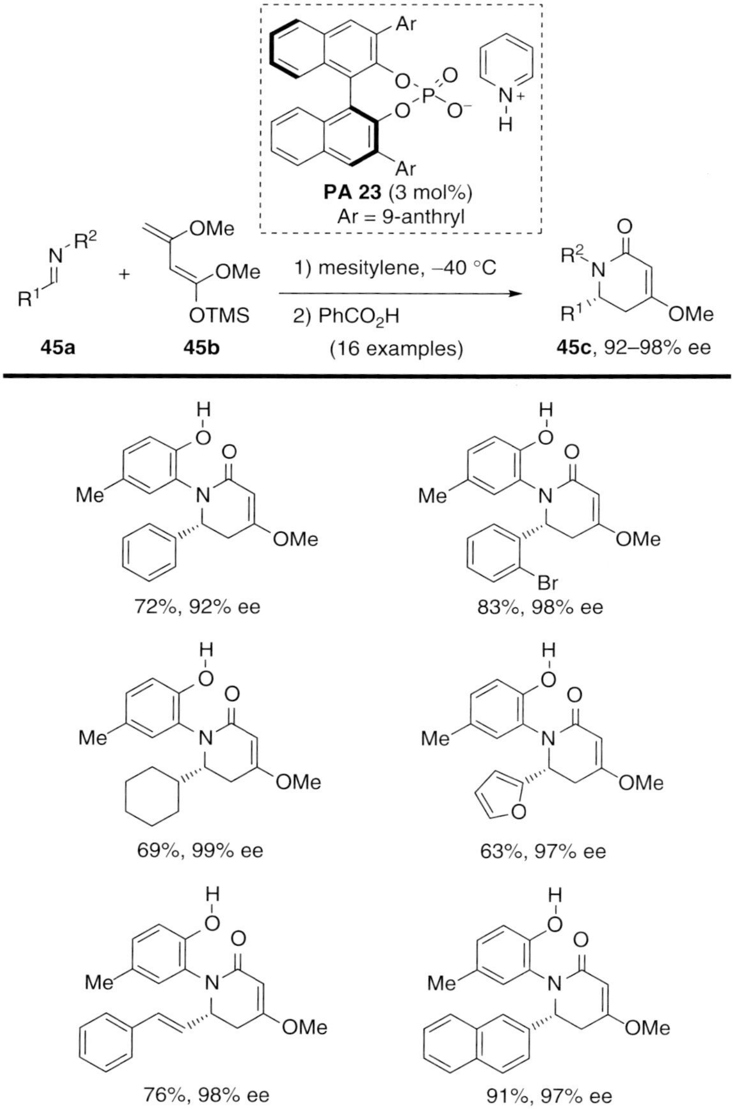

Scheme 2.75 Aza-Diels–Alder reaction with Brassard's diene by Akiyama [191].

The reaction delivers piperidinones **45c** in good yields and with excellent selectivities. The protocol utilizes a unique pyridine salt catalyst **PA 23**, whereby the pyridine is thought to be crucial in stabilizing diene **45b**. The mechanism of this process is proposed in Scheme 2.76. Activation of the imine occurs by the catalyst, which initiates the [4+2]-cycloaddition process to yield a hemiaminal intermediate, which under further separate acidic conditions can eliminate TMSOMe to yield the lactam products. It could, however, be envisioned that a stepwise process may also be occurring in this case.

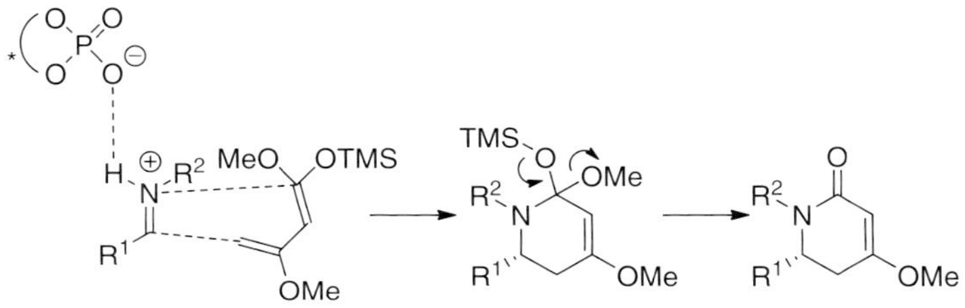

Scheme 2.76 Mechanism of aza-Diels–Alder reaction [191].

In the same year of 2006, Rueping and Gong independently developed aza-Diels–Alder reactions of cyclic dienes with imines. Rueping combined imines **46a** with cyclohexenone **46b** in the presence of catalyst **PA 22** and AcOH to access isoquinuclidines **46c** with good enantioselectivities (Scheme 2.77) [193]. A harmonious relationship was established by the two acid catalysts whereby **PA 22** was responsible for activating the imine while achiral AcOH's role was to tautomerize **46a** into its enol form.

Gong employed a similar strategy but instead solely used catalyst **[H$_8$]-PA 24** with greatly extended reaction times to deliver **46d** also with modest selectivities (Scheme 2.77, bottom) [194]. The Rueping group has also produced an aza-Diels–Alder reaction of cyclic imines with cyclopentadiene [195].

In 2006, Akiyama disclosed an inverse electron-demand aza-Diels–Alder reaction of enol ethers **47b** with aromatic imines **47a** using **PA 1** to approach tetrahydroquinolines **47c** with high levels of enantioselectivity (Scheme 2.78) [196].

A range of cyclic enol ethers could also be employed with similar enantioselectivities being achieved. Gong has shown that enol ethers derived from azlactones can also be used in a similar transformation [197]. Closely related to this protocol is the use of styrenes in place of enol ethers, and this area has received an appreciable amount of attention [198–202]. Another area that has also received considerable inspection is the use of enamines as components in aza-Diels–Alder transformations or more commonly referred to as *Povarov reactions*. Masson and Zhu kick started this field in 2009 by disclosure of such a reaction involving

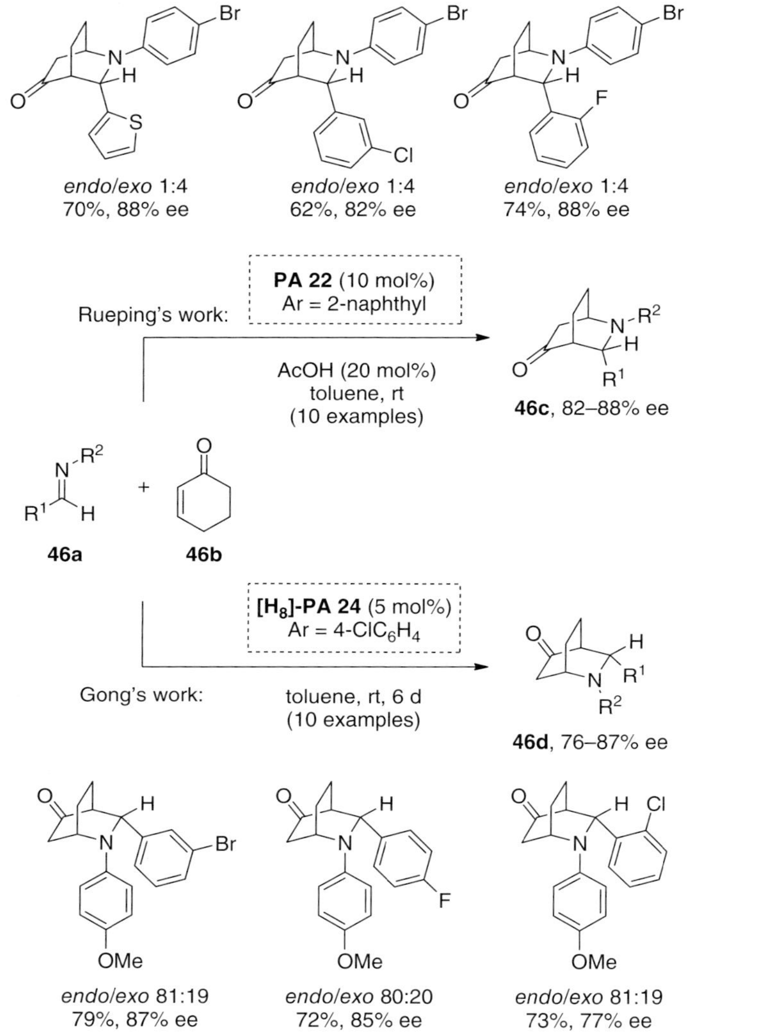

Scheme 2.77 Aza-Diels–Alder reaction with a cyclic diene by Rueping [193] and Gong [194].

aldehydes **48a**, aryl amines **48b**, and enamines **48c** (Scheme 2.79) [203–208]. Upon combination in the presence of **[H$_8$]-PA 24** delivered 2,4-substituted tetrahydroquinolines **48d** in excellent enantioselectivities.

To demonstrate its potential, the methodology was applied to a short synthesis of Torcetrapib, which has been a phase-3 drug candidate. This work has also inspired other research groups to investigate accompanying processes [209–213].

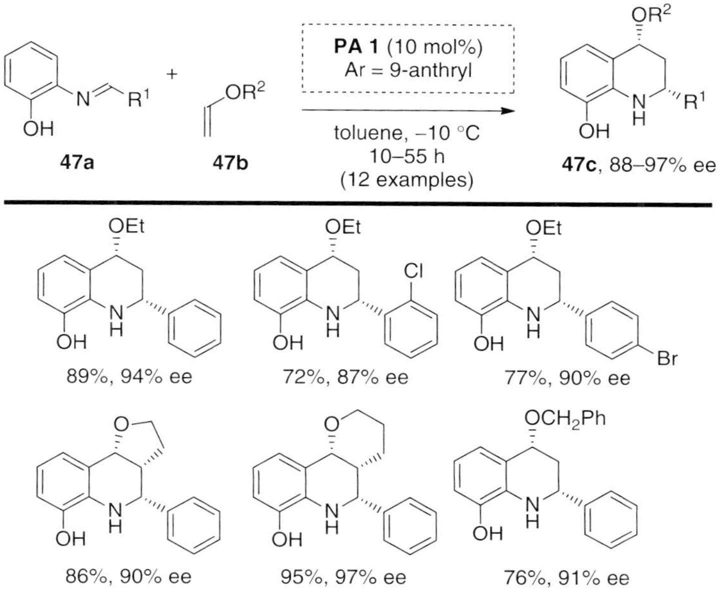

Scheme 2.78 Aza-Diels–Alder reaction with enol ethers by Akiyama [196].

An elaborate application of the aza-Diels–Alder reaction has been shown by Zhu and Sun, who found, in particular, oxetane-containing aldehydes **49b** to be the most optimal substrates for a highly complex cascade process (Scheme 2.80) [214]. Combining **49b** with aryl amines **49a** and indoles **49c** resulted in an aza-Diels–Alder reaction followed by oxetane opening to deliver the exquisite products **49d** with good enantioselectivities.

The oxetane's role is thought to involve coordination to the catalyst by acting as a hydrogen-bond acceptor to lower the activation barrier for the desired process to occur. Indeed, in its absence, the reaction yields several side products and less than 20% of the desired product. The mechanism for this elaborate process is shown in Scheme 2.81. Condensation of the aldehyde and the aniline first leads to an imine intermediate. This substrate undergoes a [4+2]-cycloaddition with the indole component. As with most of the examples within this book, it could also be a stepwise process. After the aza-Diels–Alder step, the catalyst then activates the oxetane group for a desymmetrizing ring-opening mediated by the newly formed amine to yield the observed products.

Multicomponent reactions have received considerable attention in recent years as they offer a route to preparing highly diverse libraries of compounds from simple starting materials usually with high levels of atom economy. In 2013, Lin developed the first pseudo five-component coupling reaction of aromatic aldehydes **50a**, amines **50b**, and β-keto esters **50c** (Scheme 2.82) [215].

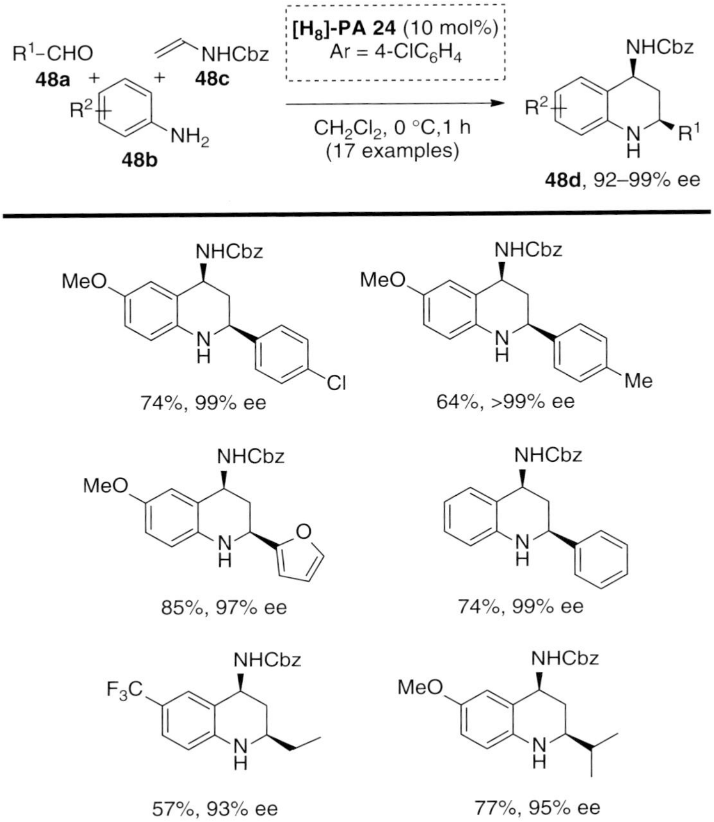

Scheme 2.79 Aza-Diels–Alder reaction of imines with enamines by Masson and Zhu [203].

The condensation consists of an AB_2C_2 type coupling and affords the tetrahydropyridines **50d** with generally high levels of enantioselectivity. Screening of various catalysts revealed SPINOL-derived **(*S*)-SPA 2** to be the optimal choice.

The exquisite process begins by a series of condensations (Scheme 2.83). Firstly, 1 equiv. of amine condenses with the β-keto ester to yield an enamine, which can go on to condense with 1 equiv. of aldehyde to yield a diene system. Meanwhile, separately, 1 equiv. of amine and 1 equiv. of aldehyde can also condense to yield an imine. The imine and the diene components are now ready to undergo a [4+2]-cycloaddition promoted by the catalyst. The authors in fact suggest that the process is a stepwise one. A comparably study has also been carried out by Shi and Tu using a BINOL-derived phosphoric acid catalyst [216].

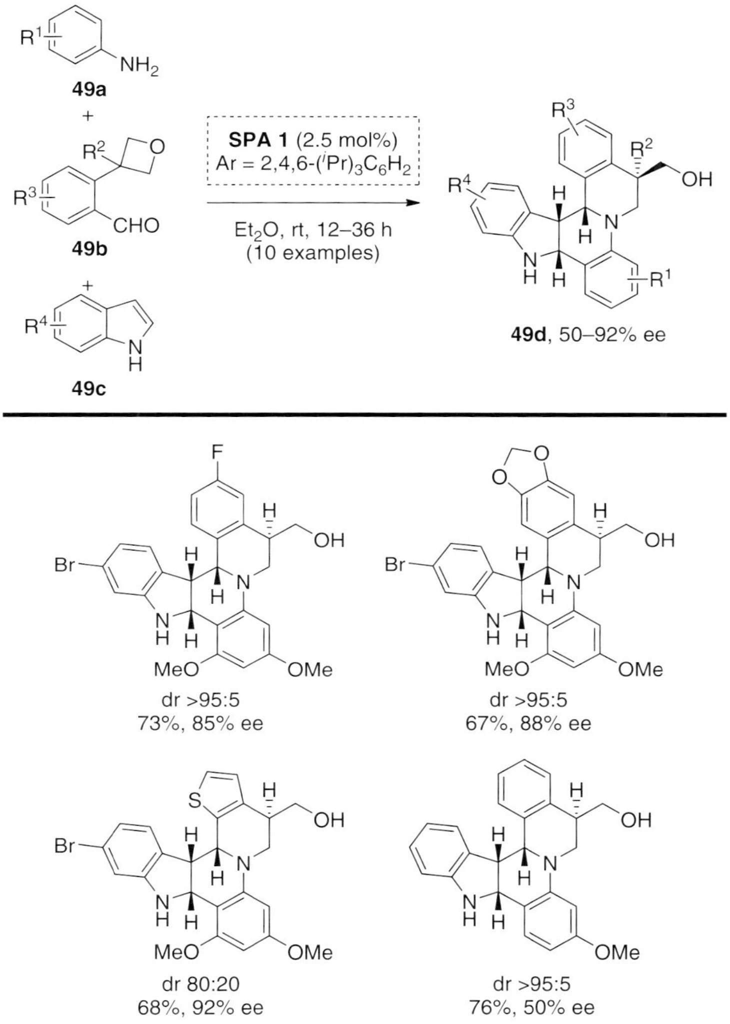

Scheme 2.80 Multicomponent aza-Diels–Alder reaction by Zhu and Sun [214].

2.7.2
1,3-Dipolar Cycloaddition

1,3-Dipolar cycloadditions have received an incredible level of attention from chemists wishing to employ chiral Brønsted acids as catalysts. The interaction between the catalyst and two opposite charges (or dipoles) in the same molecule is key to obtaining high levels of selectivity. For this duty, phosphoric acids have proved to be highly suitable candidates since they possess both a Brønsted acidic

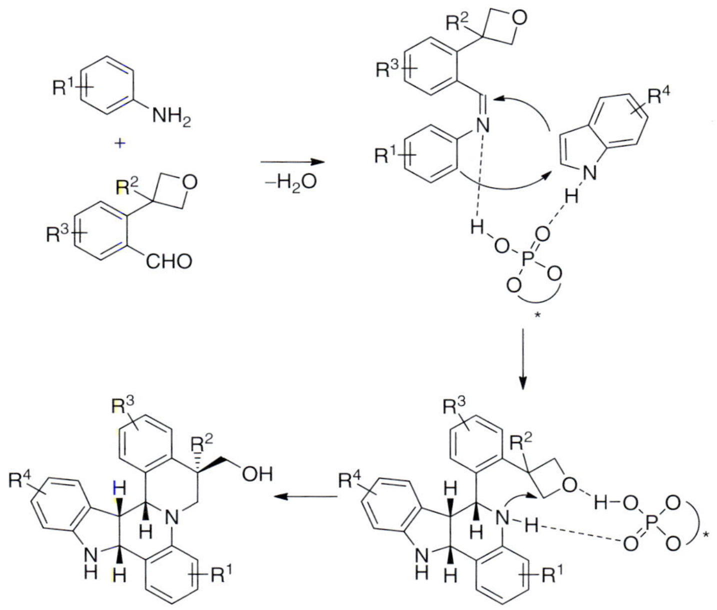

Scheme 2.81 Mechanism of multicomponent aza-Diels–Alder reaction [214].

site and a Lewis basic site. The seminal report in this area was published by Gong in 2008, who showed that maleates **51a** could react with imines formed from **51b** and **51c** to give chiral pyrrolidines **51d** with excellent enantioselectivities (Scheme 2.84) [217].

The group found that bisphosphoric acid **PA 21** delivered far more superior results than classical phosphoric acid catalysts and that the reaction performed at its best in chlorinated solvents. The mechanism for the reaction is depicted in Scheme 2.85. Condensation of the aldehyde and amine components leads to an imine. Since R^2 or R^3 is an electron-withdrawing group, the α-proton becomes acidic, and upon interaction with the catalyst, a 1,3-dipole can be established. This can then go on to react with the maleate to yield the products; however, the mechanism can be seen as general for any unsaturated component being used.

The Gong group has thoroughly explored the scope of these cyclizations by the utilization of various alternative imine and maleate combinations [218–223]. It has also inspired Gong and others [224] to replace maleates for a variety of other partners, which include nitroalkenes [225], alkynes [226–229], allenes [230, 231], and imines [232, 233]. In addition, Gong has also shown an intramolecular variant [234].

Hydrazones are well known for their ability to undergo 1,3-dipolar cycloadditions usually with alkene coupling partners. Tsogoeva communicated the seminal results on this transformation, which consisted of the use of a chiral phosphoric acid in combination with a Lewis acidic silicon additive but only delivered modest

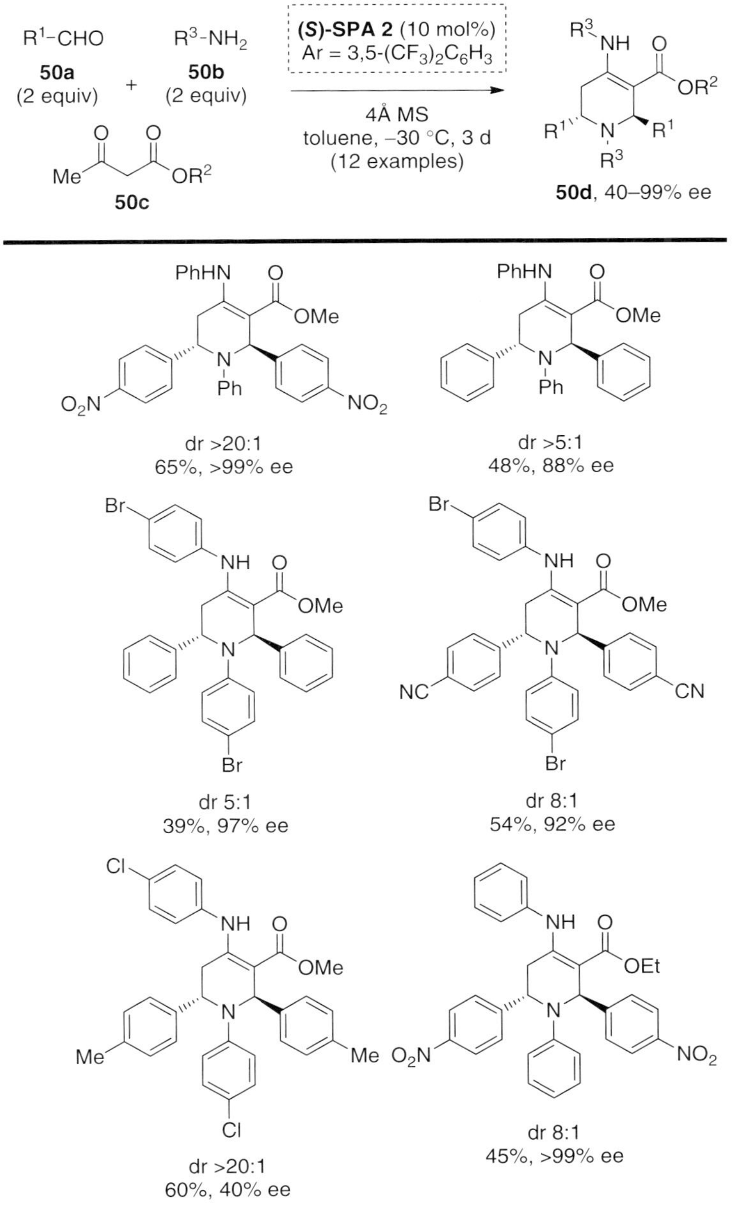

Scheme 2.82 Five-component aza-Diels–Alder type reaction by Lin [215].

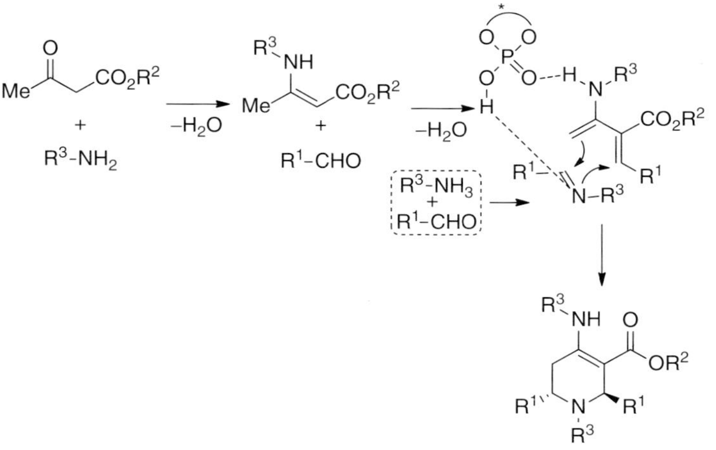

Scheme 2.83 Mechanism of five-component aza-Diels–Alder type reaction [215].

selectivities [235, 236]. In 2012, Rueping was able to develop a protocol using an *N*-triflyl phosphoramide catalyst **[H$_8$]-NPA 4**, which excluded the use of a Lewis acidic additive. They showed that hydrazones **52a** could react with cyclopentadiene **52b** or stryenes **52d** to give the corresponding products **52c** and **52e** with high to excellent levels of enantioselectivity (Scheme 2.86) [237].

The more acidic catalyst choice is proposed to allow for the absence of an additional Lewis acid since it is able to completely activate the hydrazone by itself. The pyrazolidine products were also shown to be highly suitable for a variety of further transformations without erosion of enantiopurity.

The underutilized components in 1,3-dipolar cycloadditions are cyclic azomethine imines, which offer unique opportunities to access unusual chemical motifs. In line with Maruoka's expertise, he developed a dicarboxylic acid catalyst **BA 11**, which was able to catalyze the cycloaddition between azomethine imines **53b** with vinyl ether **53a** to yield the corresponding adducts **53c** with excellent levels of enantioselectivity (Scheme 2.87) [238].

In general, the reaction proceeds exceedingly with a variety of substitution patterns tolerated on the azomethine imine. The products **53c** are all obtained from an *exo*-transition state. The reaction could also be extended to the use of vinylogous aza-enamines, which provided slightly lower levels of selectivity. The groups of Shi [239] and Hong and Wang [240] have also developed the use of cyclic azomethine imines with chiral phosphoric acid catalysts.

2.7.3
Electrocyclizations and Sigmatropic Rearrangements

Electrocyclic reactions are valuable transformations that usually unlock routes to molecules that would not be able to be accessed by conventional methods.

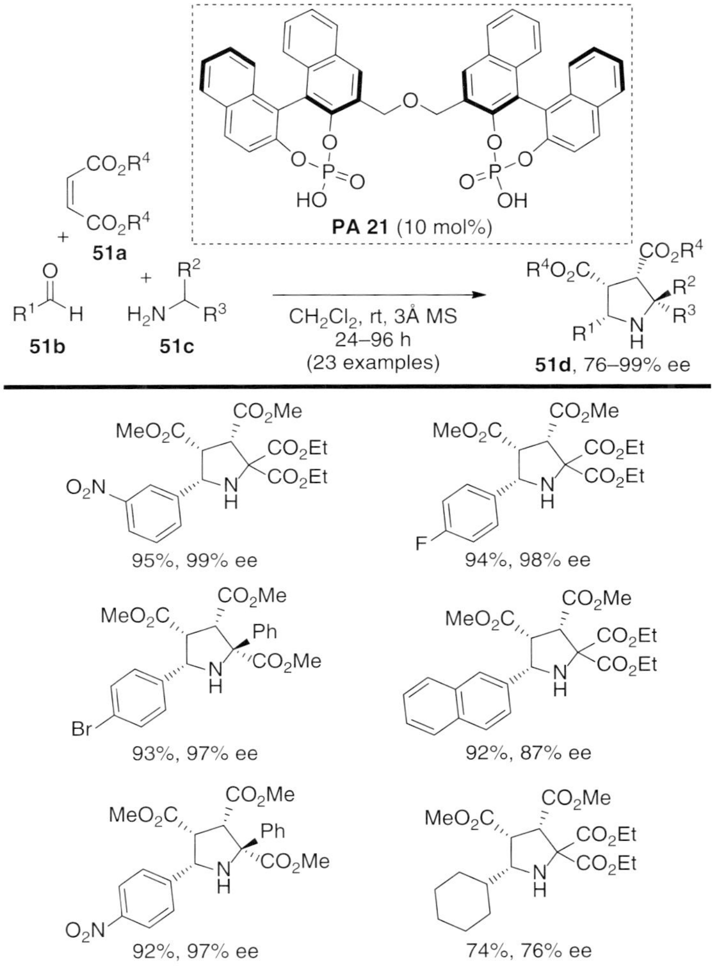

Scheme 2.84 1,3-Dipolar cycloaddition of imines with maleates by Gong [217].

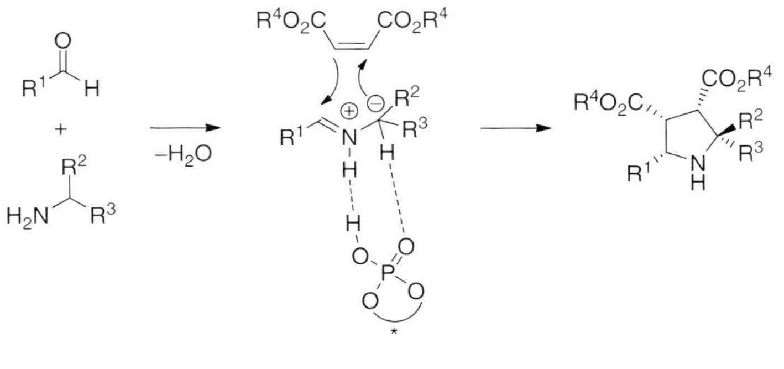

Scheme 2.85 Mechanism of 1,3-dipolar cycloaddition [217].

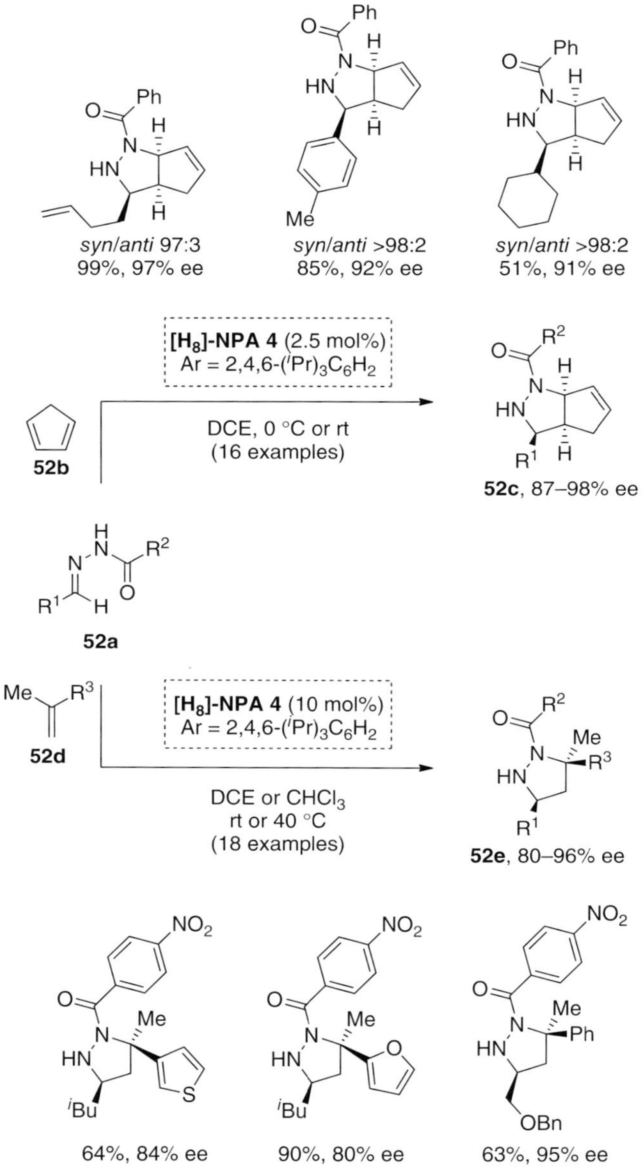

Scheme 2.86 [3+2]-Cycloadditions of hydrazones with alkenes by Rueping [237].

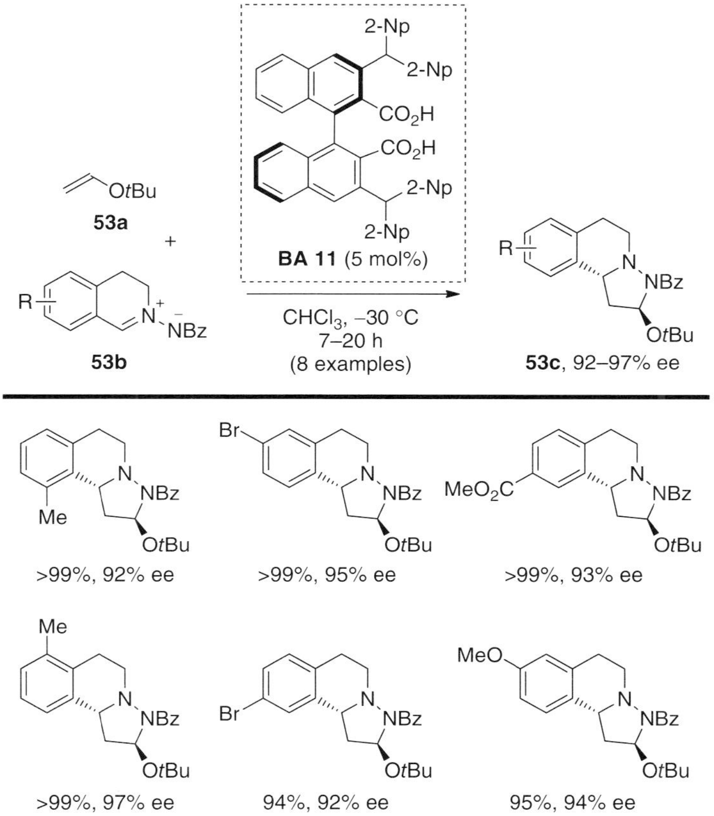

Scheme 2.87 1,3-Dipolar cycloaddition of cyclic azomethine imines by Maruoka [238].

Although cycloadditions have seen significant advances, electrocyclizations are notably more difficult to perform especially in an enantioselective manner. One of the first examples was shown by List in 2009, who demonstrated the use of hydrazones **54a** as substrates that could undergo cycloisomerization to yield pyrazolines **54b** with high enantioselectivity (Scheme 2.88) [241, 242].

A varied scope was explored for the pyrazoline products, which included both alkyl and aromatic substituents at the chiral center. In addition, the hydrazones were shown to be suitable for *in situ* generation from the corresponding hydrazine and aldehyde components. The catalyst ((*S*)-**PA 1**) is not only thought to catalyze the asymmetric transformation but is also required to isomerize the double bond geometry from *E* to *Z*.

The mechanism is given in Scheme 2.89. For the cyclization to begin, an *E* to *Z* isomerization occurs first. This Z-intermediate is then activated by the catalyst and can undergo the desired 6π-electrocyclization, which gives the product after

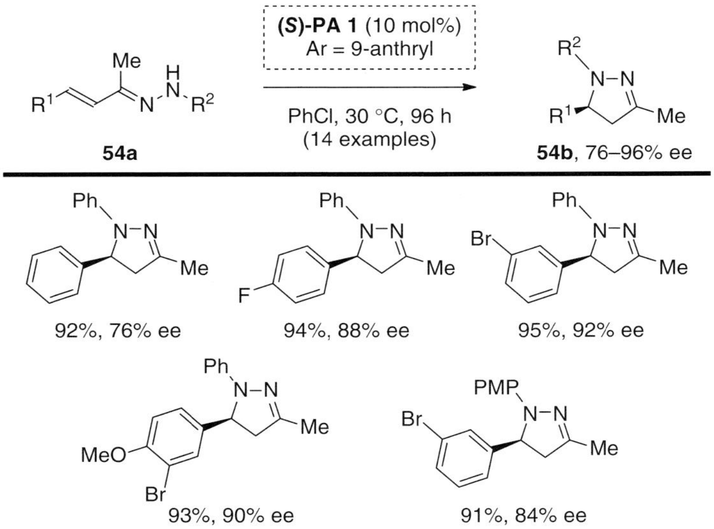

Scheme 2.88 Electrocyclizations of hydrazones by List [241].

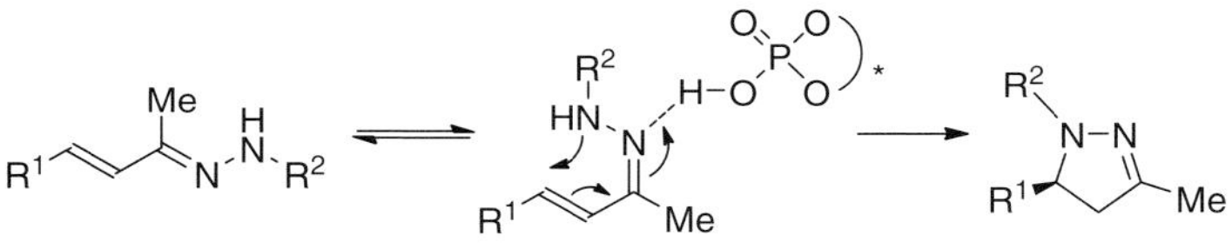

Scheme 2.89 Mechanism of electrocyclizations of hydrazones [241].

a proton transfer back to the catalyst. A conceptually related transformation to access 1,4-dihydropyridazines was also shown by Rueping in 2012 (Scheme 2.90) [243].

Coupling of fluorinated hydrazines **55a** with α,β-unsaturated aldehydes **55b** in the presence of **PA 1** gave the corresponding 1,4-dihydropyridazines **55c** in good yields and with high levels of enantioselectivity. Mechanistically, the first step involves condensation of **55a** and **55b** to give a cationic intermediate, which contains a phosphate anion, whose role is to control the torquoselectivity of the subsequent 6π-electrocyclization.

Sigmatropic rearrangements are scarcely encountered as transformations that are efficiently catalyzed by chiral Brønsted acids. The aza-Cope rearrangement is one variant that is particularly known to be difficult to control, even by other means. The first asymmetric protocol for the reaction was reported by Rueping in 2008, who utilized **[H$_8$]-PA 22** to perform the rearrangement of amine **56a** and aldehydes **56b** (Scheme 2.91) [244].

The first step of the process involves a condensation reaction, which is followed by a rearrangement process to yield optically active homoallylic amines **56c** with excellent enantiopurities. Interestingly, during the catalyst optimizations, it was

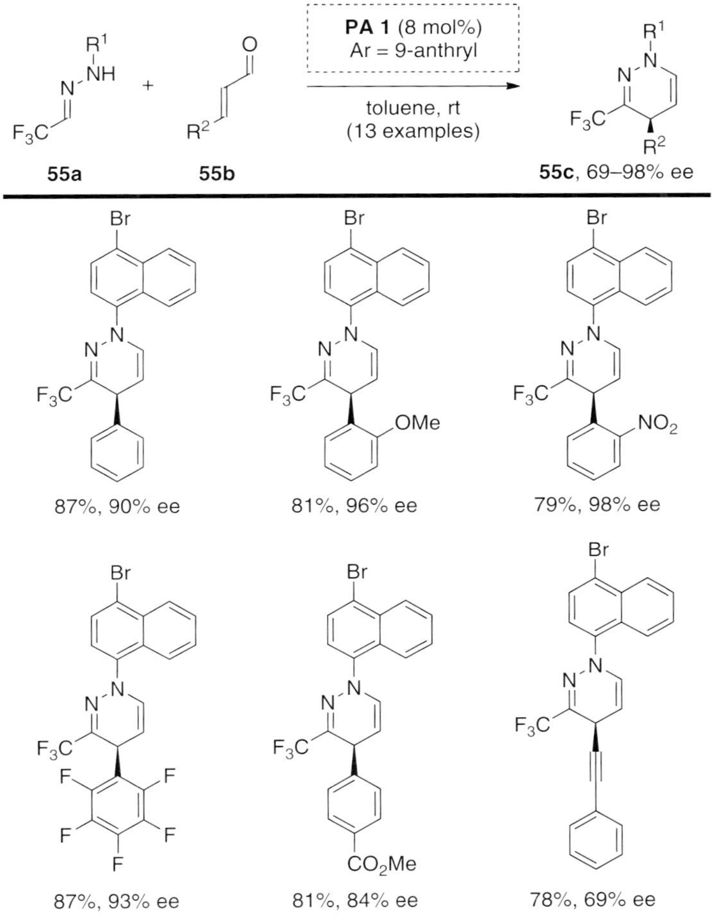

Scheme 2.90 Electrocyclizations of hydrazones by Rueping [243].

found that the octahydro version of **PA 22** delivered considerably higher levels of selectivity than the corresponding unsaturated parent catalyst.

The mechanism of the rearrangement is illustrated in Scheme 2.92. Condensation of the homoallylic amine and the aldehydes occurs first, which provides the key intermediate. Activation by the catalyst sets up the sigmatropic process, and after a proton transfer back to the catalyst, the desired product is formed.

2.8 Radical Reactions

One area of organic chemistry that has remained largely untouched with respect to chiral Brønsted acid catalysis is the field of radical reactions. Although a wealth of

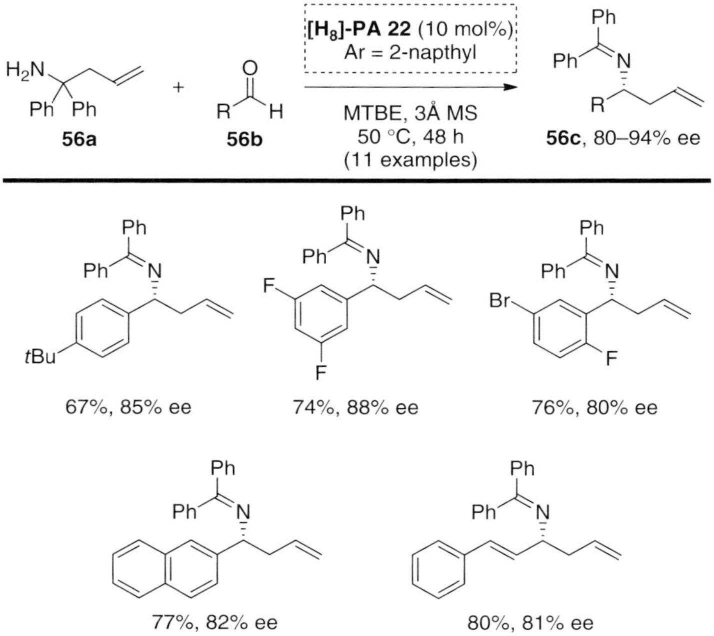

Scheme 2.91 Aza-Cope sigmatropic rearrangement by Rueping [244].

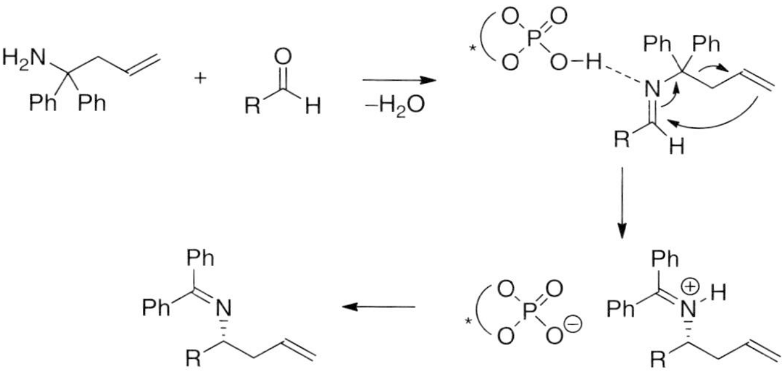

Scheme 2.92 Mechanism of aza-Cope rearrangement [244].

useful and highly exquisite transformations is known to the synthetic community, the use of Brønsted acids has not found suitable application. A rare example was reported in 2009 by Kim, who disclosed the coupling of aryl-protected imines **57a** with alkyl iodides **57b** in the presence of **NPA 4** (Scheme 2.93) [245].

The radicals were generated from the alkyl iodides **57b** by using tris(trimethylsilyl)silane (TTMSSH) and Et_3B as an initiator. The scope of the process was rather limited, and only modest yields and selectivities of the products **57c** were obtained. The main side reaction involved the addition of an ethyl group (formed

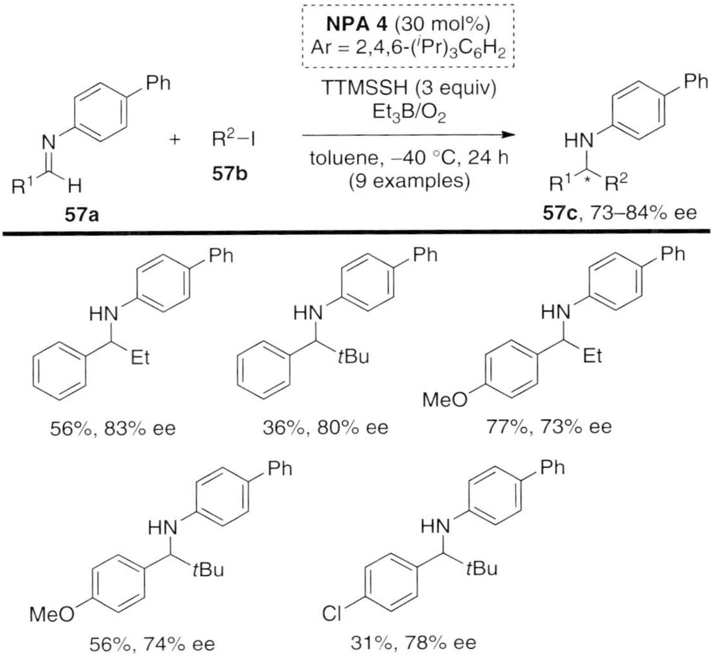

Scheme 2.93 The coupling of radicals with imines by Kim [245].

from Et_3B) to the imines. Although modest results were obtained, this report is still a breakthrough, but further studies are surely required for this area's advancement.

References

1. Uraguchi, D., Sorimachi, K., and Terada, M. (2005) *J. Am. Chem. Soc.*, **127**, 9360–9361.
2. Akiyama, T., Itoh, J., Yokota, K., and Fuchibe, K. (2004) *Angew. Chem. Int. Ed.*, **43**, 1566–1568.
3. Zhang, H., Wen, X., Gan, L., and Peng, Y. (2012) *Org. Lett.*, **14**, 2126–2129.
4. Hashimoto, T. and Maruoka, K. (2007) *J. Am. Chem. Soc.*, **129**, 10054–10055.
5. Hashimoto, T. and Maruoka, K. (2008) *Synthesis*, **2008**, 3703–3706.
6. Hashimoto, T., Kimura, H., Nakatsu, H., and Maruoka, K. (2011) *J. Org. Chem.*, **76**, 6030–6037.
7. Hashimoto, T., Kimura, H., Kawamata, Y., and Maruoka, K. (2011) *Nat. Chem.*, **3**, 642–646.
8. Hashimoto, T., Uchiyama, N., and Maruoka, K. (2008) *J. Am. Chem. Soc.*, **130**, 14380–14381.
9. Hashimoto, T., Nakatsu, H., Yamamoto, K., and Maruoka, K. (2011) *J. Am. Chem. Soc.*, **133**, 9730–9733.
10. Akiyama, T., Suzuki, T., and Mori, K. (2009) *Org. Lett.*, **11**, 2445–2447.
11. Zeng, X., Zeng, X., Xu, Z., Lu, M., and Zhong, G. (2009) *Org. Lett.*, **11**, 3036–3039.
12. Terada, M., Machioka, K., and Sorimachi, K. (2006) *Angew. Chem. Int. Ed.*, **45**, 2254–2257.
13. Terada, M., Machioka, K., and Sorimachi, K. (2007) *J. Am. Chem. Soc.*, **129**, 10336–10337.

14. Dagousset, G., Drouet, F., Masson, G., and Zhu, J. (2009) *Org. Lett.*, **11**, 5546–5549.
15. Ren, L., Lei, T., and Gong, L.-Z. (2011) *Chem. Commun.*, **47**, 11683–11685.
16. Wang, Y., Shi, F., Yao, X.-X., Sun, M., Dong, L., and Tu, S.-J. (2014) *Chem. Eur. J.*, **20**, 15047–15052.
17. Wang, Y., Tu, M.-S., Shi, F., and Tu, S.-J. (2014) *Adv. Synth. Catal.*, **356**, 2009–2019.
18. Momiyama, N., Nishimoto, H., and Terada, M. (2011) *Org. Lett.*, **13**, 2126–2129.
19. Gandhi, S. and List, B. (2013) *Angew. Chem. Int. Ed.*, **52**, 2573–2576.
20. Rueping, M., Sugiono, E., Theissmann, T., Kuenkel, A., Köckritz, A., Pews-Davtyan, A., Nemati, N., and Beller, M. (2007) *Org. Lett.*, **9**, 1065–1068.
21. Hashimoto, T., Hirose, M., and Maruoka, K. (2008) *J. Am. Chem. Soc.*, **130**, 7556–7557.
22. Hashimoto, T., Kimura, H., and Maruoka, K. (2010) *Angew. Chem. Int. Ed.*, **49**, 6844–6847.
23. Yue, T., Wang, M.-X., Wang, D.-X., Masson, G., and Zhu, J. (2009) *Angew. Chem. Int. Ed.*, **48**, 6717–6721.
24. Su, Y., Bouma, M.J., Alcaraz, L., Stocks, M., Furber, M., Masson, G., and Zhu, J. (2012) *Chem. Eur. J.*, **18**, 12624–12627.
25. Hashimoto, T., Kimura, H., Kawamata, Y., and Maruoka, K. (2012) *Angew. Chem. Int. Ed.*, **51**, 7279–7281.
26. Li, G., Kaplan, M.J., Wojtas, L., and Antilla, J.C. (2010) *Org. Lett.*, **12**, 1960–1963.
27. Rueping, M. and Lin, M.-Y. (2010) *Chem. Eur. J.*, **16**, 4169–4172.
28. Bernardi, L., Comes-Franchini, M., Fochi, M., Leo, V., Mazzanti, A., and Ricci, A. (2010) *Adv. Synth. Catal.*, **352**, 3399–3406.
29. Rueping, M. and Antonchick, A.P. (2008) *Org. Lett.*, **10**, 1731–1734.
30. Rowland, G.B., Zhang, H., Rowland, E.B., Chennamadhavuni, S., Wang, Y., and Antilla, J.C. (2005) *J. Am. Chem. Soc.*, **127**, 15696–15697.
31. Liang, Y., Rowland, E.B., Rowland, G.B., Perman, J.A., and Antilla, J.C. (2007) *Chem. Commun.*, **2007**, 4477–4479.
32. Hatano, M., Ozaki, T., Sugiura, Y., and Ishihara, K. (2012) *Chem. Commun.*, **48**, 4986–4988.
33. Cheng, X., Vellalath, S., Goddard, R., and List, B. (2008) *J. Am. Chem. Soc.*, **130**, 15786–15787.
34. Rueping, M., Antonchick, A.P., Sugiono, E., and Grenader, K. (2009) *Angew. Chem. Int. Ed.*, **48**, 908–910.
35. Cheng, D.-J., Tian, Y., and Tian, S.-K. (2012) *Adv. Synth. Catal.*, **354**, 995–999.
36. He, Y., Cheng, C., Chen, B., Duan, K., Zhuang, Y., Yuan, B., Zhang, M., Zhou, Y., Zhou, Z., Su, Y.-J., Cao, R., and Qiu, L. (2014) *Org. Lett.*, **16**, 6366–6369.
37. Jiang, Y., Liu, Y., Tu, S.-J., and Shi, F. (2013) *Tetrahedron: Asymmetry*, **24**, 1286–1296.
38. Huang, D., Li, X., Xu, F., Li, L., and Lin, X. (2013) *ACS Catal.*, **3**, 2244–2247.
39. Li, G., Fronczek, F.R., and Antilla, J.C. (2008) *J. Am. Chem. Soc.*, **130**, 12216–12217.
40. Zheng, W., Wojtas, L., and Antilla, J.C. (2010) *Angew. Chem. Int. Ed.*, **49**, 6589–6591.
41. Ingle, G.K., Mormino, M.G., Wojtas, L., and Antilla, J.C. (2011) *Org. Lett.*, **13**, 4822–4825.
42. Lu, G. and Birman, V.B. (2010) *Org. Lett.*, **13**, 356–358.
43. Wang, C., Luo, H.-W., and Gong, L.-Z. (2011) *Synlett*, **2011**, 992–994.
44. Vellalath, S., Čorić, I., and List, B. (2010) *Angew. Chem. Int. Ed.*, **49**, 9749–9752.
45. Akiyama, T., Morita, H., Itoh, J., and Fuchibe, K. (2005) *Org. Lett.*, **7**, 2583–2585.
46. Cheng, X., Goddard, R., Buth, G., and List, B. (2008) *Angew. Chem. Int. Ed.*, **47**, 5079–5081.
47. Thorat, P.B., Goswami, S.V., Magar, R.L., Patil, B.R., and Bhusare, S.R. (2013) *Eur. J. Org. Chem.*, **2013**, 5509–5516.
48. Zhao, Y., Li, X., Mo, F., Li, L., and Lin, X. (2013) *RSC Adv.*, **3**, 11895–11901.
49. Bhadury, P.S., Zhang, Y., Zhang, S., Song, B., Yang, S., Hu, D., Chen, Z., Xue, W., and Jin, L. (2009) *Chirality*, **21**, 547–557.

50. Itoh, J., Fuchibe, K., and Akiyama, T. (2008) *Synthesis*, (8), 1319–1322.
51. Yamanaka, M., Itoh, J., Fuchibe, K., and Akiyama, T. (2007) *J. Am. Chem. Soc.*, **129**, 6756–6764.
52. Uraguchi, D. and Terada, M. (2004) *J. Am. Chem. Soc.*, **126**, 5356–5357.
53. Simón, L. and Goodman, J.M. (2011) *J. Org. Chem.*, **76**, 1775–1788.
54. Guo, Q.-X., Liu, H., Guo, C., Luo, S.-W., Gu, Y., and Gong, L.-Z. (2007) *J. Am. Chem. Soc.*, **129**, 3790–3791.
55. Rueping, M., Sugiono, E., and Schoepke, F.R. (2007) *Synlett*, (9), 1441–1445.
56. Ávila, E.P., Justo, R.M.S., Gonçalves, V.P., Pereira, A.A., Diniz, R., and Amarante, G.W. (2015) *J. Org. Chem.*, **80**, 590–594.
57. Hatano, M., Maki, T., Moriyama, K., Arinobe, M., and Ishihara, K. (2008) *J. Am. Chem. Soc.*, **130**, 16858–16860.
58. Terada, M., Sorimachi, K., and Uraguchi, D. (2006) *Synlett*, (1), 133–136.
59. Akiyama, T., Saitoh, Y., Morita, H., and Fuchibe, K. (2005) *Adv. Synth. Catal.*, **347**, 1523–1526.
60. Akiyama, T., Katoh, T., Mori, K., and Kanno, K. (2009) *Synlett*, (10), 1664–1666.
61. Kashikura, W., Mori, K., and Akiyama, T. (2011) *Org. Lett.*, **13**, 1860–1863.
62. Chen, Y.-Y., Jiang, Y.-J., Fan, Y.-S., Sha, D., Wang, Q., Zhang, G., Zheng, L., and Zhang, S. (2012) *Tetrahedron: Asymmetry*, **23**, 904–909.
63. Wang, Q., Leutzsch, M., van Gemmeren, M., and List, B. (2013) *J. Am. Chem. Soc.*, **135**, 15334–15337.
64. Hasegawa, A., Naganawa, Y., Fushimi, M., Ishihara, K., and Yamamoto, H. (2006) *Org. Lett.*, **8**, 3175–3178.
65. Sickert, M. and Schneider, C. (2008) *Angew. Chem. Int. Ed.*, **47**, 3631–3634.
66. Giera, D.S., Sickert, M., and Schneider, C. (2008) *Org. Lett.*, **10**, 4259–4262.
67. Sickert, M., Abels, F., Lang, M., Sieler, J., Birkemeyer, C., and Schneider, C. (2010) *Chem. Eur. J.*, **16**, 2806–2818.
68. Abels, F. and Schneider, C. (2011) *Synthesis*, **2011**, 4050–4058.
69. Abels, F., Lindemann, C., and Schneider, C. (2014) *Chem. Eur. J.*, **20**, 1964–1979.
70. Giera, D.S., Sickert, M., and Schneider, C. (2009) *Synthesis*, **2009**, 3797–3802.
71. Ruff, B.M., Zhong, S., Nieger, M., Sickert, M., Schneider, C., and Bräse, S. (2011) *Eur. J. Org. Chem.*, **2011**, 6558–6566.
72. Akiyama, T., Honma, Y., Itoh, J., and Fuchibe, K. (2008) *Adv. Synth. Catal.*, **350**, 399–402.
73. Fritzsche, S., Ohla, S., Glaser, P., Giera, D.S., Sickert, M., Schneider, C., and Belder, D. (2011) *Angew. Chem. Int. Ed.*, **50**, 9467–9470.
74. Wang, Q., van Gemmeren, M., and List, B. (2014) *Angew. Chem. Int. Ed.*, **53**, 13592–13595.
75. Rueping, M., Sugiono, E., and Azap, C. (2006) *Angew. Chem. Int. Ed.*, **45**, 2617–2619.
76. Rueping, M., Sugiono, E., and Moreth, S.A. (2007) *Adv. Synth. Catal.*, **349**, 759–764.
77. Barbero, M., Cadamuro, S., Dughera, S., and Torregrossa, R. (2014) *Org. Biomol. Chem.*, **12**, 3902–3911.
78. Simón, L. and Goodman, J.M. (2009) *J. Am. Chem. Soc.*, **131**, 4070–4077.
79. Zamfir, A. and Tsogoeva, S.B. (2009) *Org. Lett.*, **12**, 188–191.
80. Zhang, G.-W., Zheng, D.-H., Nie, J., Wang, T., and Ma, J.-A. (2010) *Org. Biomol. Chem.*, **8**, 1399–1405.
81. Hashimoto, T., Kimura, H., and Maruoka, K. (2010) *Tetrahedron: Asymmetry*, **21**, 1187–1188.
82. Chen, X.-H., Xu, X.-Y., Liu, H., Cun, L.-F., and Gong, L.-Z. (2006) *J. Am. Chem. Soc.*, **128**, 14802–14803.
83. Li, N., Chen, X.-H., Song, J., Luo, S.-W., Fan, W., and Gong, L.-Z. (2009) *J. Am. Chem. Soc.*, **131**, 15301–15310.
84. Li, N., Chen, X.-H., Zhou, S.-M., Luo, S.-W., Song, J., Ren, L., and Gong, L.-Z. (2010) *Angew. Chem. Int. Ed.*, **49**, 6378–6381.
85. Xu, F., Huang, D., Lin, X., and Wang, Y. (2012) *Org. Biomol. Chem.*, **10**, 4467–4470.
86. An, D., Fan, Y.-S., Gao, Y., Zhu, Z.-Q., Zheng, L.-Y., and Zhang, S.-Q. (2014) *Eur. J. Org. Chem.*, **2014**, 301–306.

87. Kang, Q., Zhao, Z.-A., and You, S.-L. (2007) *J. Am. Chem. Soc.*, **129**, 1484–1485.
88. Kang, Q., Zhao, Z.A., and You, S.-L. (2009) *Tetrahedron*, **65**, 1603–1607.
89. Rowland, G.B., Rowland, E.B., Liang, Y., Perman, J.A., and Antilla, J.C. (2007) *Org. Lett.*, **9**, 2609–2611.
90. Terada, M., Yokoyama, S., Sorimachi, K., and Uraguchi, D. (2007) *Adv. Synth. Catal.*, **349**, 1863–1867.
91. Wanner, M.J., Hauwert, P., Schoemaker, H.E., de Gelder, R., van Maarseveen, J.H., and Hiemstra, H. (2008) *Eur. J. Org. Chem.*, **2008**, 180–185.
92. Zhang, G.-W., Wang, L., Nie, J., and Ma, J.-A. (2008) *Adv. Synth. Catal.*, **350**, 1457–1463.
93. Xu, F., Huang, D., Han, C., Shen, W., Lin, X., and Wang, Y. (2010) *J. Org. Chem.*, **75**, 8677–8680.
94. Xing, C.-H., Liao, Y.-X., Ng, J., and Hu, Q.-S. (2011) *J. Org. Chem.*, **76**, 4125–4131.
95. Osorio-Planes, L., Rodríguez-Escrich, C., and Pericàs, M.A. (2014) *Chem. Eur. J.*, **20**, 2367–2372.
96. Chen, L.-Y., He, H., Chan, W.-H., and Lee, A.W.M. (2011) *J. Org. Chem.*, **76**, 7141–7147.
97. Ube, H., Fukuchi, S., and Terada, M. (2010) *Tetrahedron: Asymmetry*, **21**, 1203–1205.
98. Wu, K., Jiang, Y.-J., Fan, Y.-S., Sha, D., and Zhang, S. (2013) *Chem. Eur. J.*, **19**, 474–478.
99. Enders, D., Narine, A.A., Toulgoat, F., and Bisschops, T. (2008) *Angew. Chem. Int. Ed.*, **47**, 5661–5665.
100. Enders, D., Liebich, J.X., and Raabe, G. (2010) *Chem. Eur. J.*, **16**, 9763–9766.
101. Simón, L. and Goodman, J.M. (2010) *J. Org. Chem.*, **75**, 589–597.
102. Rueping, M., Raja, S., and Núñez, A. (2011) *Adv. Synth. Catal.*, **353**, 563–568.
103. Zhang, K.-F., Nie, J., Guo, R., Zheng, Y., and Ma, J.-A. (2013) *Adv. Synth. Catal.*, **355**, 3497–3502.
104. Wang, S.-G., Zhang, W., and You, S.-L. (2013) *Org. Lett.*, **15**, 1488–1491.
105. Kano, T., Takechi, R., Kobayashi, R., and Maruoka, K. (2014) *Org. Biomol. Chem.*, **12**, 724–727.
106. Feng, J., Yan, W., Wang, D., Li, P., Sun, Q., and Wang, R. (2012) *Chem. Commun.*, **48**, 8003–8005.
107. Husmann, R., Sugiono, E., Mersmann, S., Raabe, G., Rueping, M., and Bolm, C. (2011) *Org. Lett.*, **13**, 1044–1047.
108. Qian, Y., Jing, C., Zhai, C., and Hu, W.-H. (2012) *Adv. Synth. Catal.*, **354**, 301–307.
109. Uraguchi, D., Sorimachi, K., and Terada, M. (2004) *J. Am. Chem. Soc.*, **126**, 11804–11805.
110. Li, G., Rowland, G.B., Rowland, E.B., and Antilla, J.C. (2007) *Org. Lett.*, **9**, 4065–4068.
111. Nakamura, S., Sakurai, Y., Nakashima, H., Shibata, N., and Toru, T. (2009) *Synlett*, (10), 1639–1642.
112. Kang, Q., Zheng, X., and You, S.-L. (2008) *Chem. Eur. J.*, **14**, 3539–3542.
113. Hatano, M., Sugiura, Y., Akakura, M., and Ishihara, K. (2011) *Synlett*, **2011**, 1247–1250.
114. Fan, Y.-S., Jiang, Y.-J., An, D., Sha, D., Antilla, J.C., and Zhang, S. (2014) *Org. Lett.*, **16**, 6112–6115.
115. Enders, D., Seppelt, M., and Beck, T. (2010) *Adv. Synth. Catal.*, **352**, 1413–1418.
116. Seayad, J., Seayad, A.M., and List, B. (2006) *J. Am. Chem. Soc.*, **128**, 1086–1087.
117. Wanner, M.J., van der Haas, R.N.S., de Cuba, K.R., van Maarseveen, J.H., and Hiemstra, H. (2007) *Angew. Chem. Int. Ed.*, **46**, 7485–7487.
118. Sewgobind, N.V., Wanner, M.J., Ingemann, S., de Gelder, R., van Maarseveen, J.H., and Hiemstra, H. (2008) *J. Org. Chem.*, **73**, 6405–6408.
119. Wanner, M.J., Boots, R.N.A., Eradus, B., de Gelder, R., van Maarseveen, J.H., and Hiemstra, H. (2009) *Org. Lett.*, **11**, 2579–2581.
120. Wanner, M.J., Claveau, E., van Maarseveen, J.H., and Hiemstra, H. (2011) *Chem. Eur. J.*, **17**, 13680–13683.
121. Herlé, B., Wanner, M.J., van Maarseveen, J.H., and Hiemstra, H. (2011) *J. Org. Chem.*, **76**, 8907–8912.
122. He, Y., Lin, M., Li, Z., Liang, X., Li, G., and Antilla, J.C. (2011) *Org. Lett.*, **13**, 4490–4493.

123. Rueping, M. and Volla, C.M.R. (2011) *RSC Adv.*, **1**, 79–82.
124. Cheng, D.-J., Wu, H.-B., and Tian, S.-K. (2011) *Org. Lett.*, **13**, 5636–5639.
125. Huang, D., Xu, F., Lin, X., and Wang, Y. (2012) *Chem. Eur. J.*, **18**, 3148–3152.
126. Li, X., Chen, D., Gu, H., and Lin, X. (2014) *Chem. Commun.*, **50**, 7538–7541.
127. Duce, S., Pesciaioli, F., Gramigna, L., Bernardi, L., Mazzanti, A., Ricci, A., Bartoli, G., and Bencivenni, G. (2011) *Adv. Synth. Catal.*, **353**, 860–864.
128. Badillo, J.J., Silva-García, A., Shupe, B.H., Fettinger, J.C., and Franz, A.K. (2011) *Tetrahedron Lett.*, **52**, 5550–5553.
129. Muratore, M.E., Holloway, C.A., Pilling, A.W., Storer, R.I., Trevitt, G., and Dixon, D.J. (2009) *J. Am. Chem. Soc.*, **131**, 10796–10797.
130. Holloway, C.A., Muratore, M.E., Storer, R.I., and Dixon, D.J. (2010) *Org. Lett.*, **12**, 4720–4723.
131. Muratore, M.E., Shi, L., Pilling, A.W., Storer, R.I., and Dixon, D.J. (2012) *Chem. Commun.*, **48**, 6351–6353.
132. Aillaud, I., Barber, D.M., Thompson, A.L., and Dixon, D.J. (2013) *Org. Lett.*, **15**, 2946–2949.
133. Rueping, M., Sugiono, E., Azap, C., Theissmann, T., and Bolte, M. (2005) *Org. Lett.*, **7**, 3781–3783.
134. Hoffmann, S., Seayad, A.M., and List, B. (2005) *Angew. Chem. Int. Ed.*, **44**, 7424–7427.
135. Nguyen, T.B., Bousserouel, H., Wang, Q., and Guéritte, F. (2011) *Adv. Synth. Catal.*, **353**, 257–262.
136. Nguyen, T.B., Bousserouel, H., Wang, Q., and Guéritte, F. (2010) *Org. Lett.*, **12**, 4705–4707.
137. Nguyen, T.B., Wang, Q., and Guéritte, F. (2011) *Chem. Eur. J.*, **17**, 9576–9580.
138. Storer, R.I., Carrera, D.E., Ni, Y., and MacMillan, D.W.C. (2006) *J. Am. Chem. Soc.*, **128**, 84–86.
139. Wakchaure, V.N., Nicoletti, M., Ratjen, L., and List, B. (2010) *Synlett*, **2010**, 2708–2710.
140. Hoffmann, S., Nicoletti, M., and List, B. (2006) *J. Am. Chem. Soc.*, **128**, 13074–13075.
141. Wakchaure, V.N., Zhou, J., Hoffmann, S., and List, B. (2010) *Angew. Chem. Int. Ed.*, **49**, 4612–4614.
142. Zhu, C. and Akiyama, T. (2009) *Org. Lett.*, **11**, 4180–4183.
143. Henseler, A., Kato, M., Mori, K., and Akiyama, T. (2011) *Angew. Chem. Int. Ed.*, **50**, 8180–8183.
144. Sakamoto, T., Horiguchi, K., Saito, K., Mori, K., and Akiyama, T. (2013) *Asian J. Org. Chem.*, **2**, 943–946.
145. Sakamoto, T., Mori, K., and Akiyama, T. (2012) *Org. Lett.*, **14**, 3312–3315.
146. Saito, K. and Akiyama, T. (2012) *Chem. Commun.*, **48**, 4573–4575.
147. Saito, K., Horiguchi, K., Shibata, Y., Yamanaka, M., and Akiyama, T. (2014) *Chem. Eur. J.*, **20**, 7616–7620.
148. Hachiya, I., Ito, A., and Shimizu, M. (2014) *Asian J. Org. Chem.*, **3**, 614–618.
149. Enders, D., Rembiak, A., and Seppelt, M. (2013) *Tetrahedron Lett.*, **54**, 470–473.
150. Saito, K., Miyashita, H., and Akiyama, T. (2014) *Org. Lett.*, **16**, 5312–5315.
151. Saito, K., Shibata, Y., Yamanaka, M., and Akiyama, T. (2013) *J. Am. Chem. Soc.*, **135**, 11740–11743.
152. Li, G., Liang, Y., and Antilla, J.C. (2007) *J. Am. Chem. Soc.*, **129**, 5830–5831.
153. Kang, Q., Zhao, Z.-A., and You, S.-L. (2007) *Adv. Synth. Catal.*, **349**, 1657–1660.
154. Kang, Q., Zhao, Z.-A., and You, S.-L. (2008) *Org. Lett.*, **10**, 2031–2034.
155. Zhu, C. and Akiyama, T. (2010) *Adv. Synth. Catal.*, **352**, 1846–1850.
156. Enders, D., Rembiak, A., and Stöckel, B.A. (2013) *Adv. Synth. Catal.*, **355**, 1937–1942.
157. Rueping, M., Antonchick, A.P., and Theissmann, T. (2006) *Angew. Chem. Int. Ed.*, **45**, 3683–3686.
158. Rueping, M. and Hubener, L. (2011) *Synlett*, **2011**, 1243–1246.
159. Rueping, M., Theissmann, T., Stoeckel, M., and Antonchick, A.P. (2011) *Org. Biomol. Chem.*, **9**, 6844–6850.
160. Rueping, M. and Theissmann, T. (2010) *Chem. Sci.*, **1**, 473–476.
161. Rueping, M., Bootwicha, T., and Sugiono, E. (2012) *Beilstein J. Org. Chem.*, **8**, 300–307.

162. Guo, Q.-S., Du, D.-M., and Xu, J. (2008) *Angew. Chem. Int. Ed.*, **47**, 759–762.
163. Ferry, A., Stemper, J., Marinetti, A., Voituriez, A., and Guinchard, X. (2014) *Eur. J. Org. Chem.*, **2014**, 188–193.
164. Rueping, M., Sugiono, E., Steck, A., and Theissmann, T. (2010) *Adv. Synth. Catal.*, **352**, 281–287.
165. Bleschke, C., Schmidt, J., Kundu, D.S., Blechert, S., and Thomas, A. (2011) *Adv. Synth. Catal.*, **353**, 3101–3106.
166. Kundu, D.S., Schmidt, J., Bleschke, C., Thomas, A., and Blechert, S. (2012) *Angew. Chem. Int. Ed.*, **51**, 5456–5459.
167. Mutyala, A.K. and Patil, N.T. (2014) *Org. Chem. Front.*, **1**, 582–586.
168. Stemper, J., Isaac, K., Duret, V., Retailleau, P., Voituriez, A., Betzer, J.-F., and Marinetti, A. (2013) *Chem. Commun.*, **49**, 6084–6086.
169. Stemper, J., Isaac, K., Pastor, J., Frison, G., Retailleau, P., Voituriez, A., Betzer, J.-F., and Marinetti, A. (2013) *Adv. Synth. Catal.*, **355**, 3613–3624.
170. Isaac, K., Stemper, J., Servajean, V., Retailleau, P., Pastor, J., Frison, G., Kaupmees, K., Leito, I., Betzer, J.-F., and Marinetti, A. (2014) *J. Org. Chem.*, **79**, 9639–9646.
171. Chen, M.-W., Cai, X.-F., Chen, Z.-P., Shi, L., and Zhou, Y.-G. (2014) *Chem. Commun.*, **50**, 12526–12529.
172. Aillerie, A., de Talancé, V.L., Moncomble, A., Bousquet, T., and Pélinski, L. (2014) *Org. Lett.*, **16**, 2982–2985.
173. Cai, X.-F., Guo, R.-N., Feng, G.-S., Wu, B., and Zhou, Y.-G. (2014) *Org. Lett.*, **16**, 2680–2683.
174. Metallinos, C., Barrett, F.B., and Xu, S. (2008) *Synlett*, (5), 720–724.
175. Rueping, M., Tato, F., and Schoepke, F.R. (2010) *Chem. Eur. J.*, **16**, 2688–2691.
176. Shi, F., Tan, W., Zhang, H.-H., Li, M., Ye, Q., Ma, G.-H., Tu, S.-J., and Li, G. (2013) *Adv. Synth. Catal.*, **355**, 3715–3726.
177. Rueping, M., Antonchick, A.P., and Theissmann, T. (2006) *Angew. Chem. Int. Ed.*, **45**, 6751–6755.
178. Rueping, M., Stoeckel, M., Sugiono, E., and Theissmann, T. (2010) *Tetrahedron*, **66**, 6565–6568.
179. Rueping, M. and Antonchick, A.P. (2007) *Angew. Chem. Int. Ed.*, **46**, 4562–4565.
180. Han, Z.-Y., Xiao, H., and Gong, L.-Z. (2009) *Bioorg. Med. Chem. Lett.*, **19**, 3729–3732.
181. Rueping, M., Brinkmann, C., Antonchick, A.P., and Atodiresei, I. (2010) *Org. Lett.*, **12**, 4604–4607.
182. Rueping, M., Merino, E., and Koenigs, R.M. (2010) *Adv. Synth. Catal.*, **352**, 2629–2634.
183. Zhou, J. and List, B. (2007) *J. Am. Chem. Soc.*, **129**, 7498–7499.
184. He, Y.-P., Du, Y.-L., Luo, S.-W., and Gong, L.-Z. (2011) *Tetrahedron Lett.*, **52**, 7064–7066.
185. Liao, H.-H., Hsiao, C.-C., Sugiono, E., and Rueping, M. (2013) *Chem. Commun.*, **49**, 7953–7955.
186. Sugiono, E. and Rueping, M. (2013) *Beilstein J. Org. Chem.*, **9**, 2457–2462.
187. Rueping, M., Theissmann, T., Raja, S., and Bats, J.W. (2008) *Adv. Synth. Catal.*, **350**, 1001–1006.
188. Cai, X.-F., Chen, M.-W., Ye, Z.-S., Guo, R.-N., Shi, L., Li, Y.-Q., and Zhou, Y.-G. (2013) *Chem. Asian J.*, **8**, 1381–1385.
189. Guo, R.-N., Chen, Z.-P., Cai, X.-F., and Zhou, Y.-G. (2014) *Synthesis*, **46**, 2751–2756.
190. Cheng, D.-J., Yan, L., Tian, S.-K., Wu, M.-Y., Wang, L.-X., Fan, Z.-L., Zheng, S.-C., Liu, X.-Y., and Tan, B. (2014) *Angew. Chem. Int. Ed.*, **53**, 3684–3687.
191. Itoh, J., Fuchibe, K., and Akiyama, T. (2006) *Angew. Chem. Int. Ed.*, **45**, 4796–4798.
192. Akiyama, T., Tamura, Y., Itoh, J., Morita, H., and Fuchibe, K. (2006) *Synlett*, (1), 141–143.
193. Rueping, M. and Azap, C. (2006) *Angew. Chem. Int. Ed.*, **45**, 7832–7835.
194. Liu, H., Cun, L.-F., Mi, A.-Q., Jiang, Y.-Z., and Gong, L.-Z. (2006) *Org. Lett.*, **8**, 6023–6026.
195. Rueping, M. and Raja, S. (2012) *Beilstein J. Org. Chem.*, **8**, 1819–1824.
196. Akiyama, T., Morita, H., and Fuchibe, K. (2006) *J. Am. Chem. Soc.*, **128**, 13070–13071.
197. Jiang, J., Qing, J., and Gong, L.-Z. (2009) *Chem. Eur. J.*, **15**, 7031–7034.

198. Bergonzini, G., Gramigna, L., Mazzanti, A., Fochi, M., Bernardi, L., and Ricci, A. (2010) *Chem. Commun.*, **46**, 327–329.
199. Caruana, L., Fochi, M., Ranieri, S., Mazzanti, A., and Bernardi, L. (2013) *Chem. Commun.*, **49**, 880–882.
200. He, L., Bekkaye, M., Retailleau, P., and Masson, G. (2012) *Org. Lett.*, **14**, 3158–3161.
201. Shi, F., Xing, G.-J., Tao, Z.-L., Luo, S.-W., Tu, S.-J., and Gong, L.-Z. (2012) *J. Org. Chem.*, **77**, 6970–6979.
202. Shi, F., Xing, G.-J., Zhu, R.-Y., Tan, W., and Tu, S. (2013) *Org. Lett.*, **15**, 128–131.
203. Liu, H., Dagousset, G., Masson, G., Retailleau, P., and Zhu, J. (2009) *J. Am. Chem. Soc.*, **131**, 4598–4599.
204. Dagousset, G., Zhu, J., and Masson, G. (2011) *J. Am. Chem. Soc.*, **133**, 14804–14813.
205. Dagousset, G., Retailleau, P., Masson, G., and Zhu, J. (2012) *Chem. Eur. J.*, **18**, 5869–5873.
206. He, L., Laurent, G., Retailleau, P., Folléas, B., Brayer, J.-L., and Masson, G. (2013) *Angew. Chem. Int. Ed.*, **52**, 11088–11091.
207. Brioche, J., Courant, T., Alcaraz, L., Stocks, M., Furber, M., Zhu, J., and Masson, G. (2014) *Adv. Synth. Catal.*, **356**, 1719–1724.
208. Dagousset, G., Erb, W., Zhu, J., and Masson, G. (2014) *Org. Lett.*, **16**, 2554–2557.
209. Jiang, J., Yu, J., Sun, X.-X., Rao, Q.-Q., and Gong, L.-Z. (2008) *Angew. Chem. Int. Ed.*, **47**, 2458–2462.
210. Lin, J.-H., Zong, G., Du, R.-B., Xiao, J.-C., and Liu, S. (2012) *Chem. Commun.*, **48**, 7738–7740.
211. Luo, C. and Huang, Y. (2013) *J. Am. Chem. Soc.*, **135**, 8193–8196.
212. Huang, D., Xu, F., Chen, T., Wang, Y., and Lin, X. (2013) *RSC Adv.*, **3**, 573–578.
213. Zhang, H.-H., Sun, X.-X., Liang, J., Wang, Y.-M., Zhao, C.-C., and Shi, F. (2014) *Org. Biomol. Chem.*, **12**, 9539–9546.
214. Chen, Z., Wang, B., Wang, Z., Zhu, G., and Sun, J. (2013) *Angew. Chem. Int. Ed.*, **52**, 2027–2031.
215. Li, X., Zhao, Y., Qu, H., Mao, Z., and Lin, X. (2013) *Chem. Commun.*, **49**, 1401–1403.
216. Shi, F., Tan, W., Zhu, R.-Y., Xing, G.-J., and Tu, S.-J. (2013) *Adv. Synth. Catal.*, **355**, 1605–1622.
217. Chen, X.-H., Zhang, W.-Q., and Gong, L.-Z. (2008) *J. Am. Chem. Soc.*, **130**, 5652–5653.
218. Chen, X.-H., Wei, Q., Luo, S.-W., Xiao, H., and Gong, L.-Z. (2009) *J. Am. Chem. Soc.*, **131**, 13819–13825.
219. Wang, C., Chen, X.-H., Zhou, S.-M., and Gong, L.-Z. (2010) *Chem. Commun.*, **46**, 1275–1277.
220. He, L., Chen, X.-H., Wang, D.-N., Luo, S.-W., Zhang, W.-Q., Yu, J., Ren, L., and Gong, L.-Z. (2011) *J. Am. Chem. Soc.*, **133**, 13504–13518.
221. Cheng, M.-N., Wang, H., and Gong, L.-Z. (2011) *Org. Lett.*, **13**, 2418–2421.
222. Shi, F., Tao, Z.-L., Luo, S.-W., Tu, S.-J., and Gong, L.-Z. (2012) *Chem. Eur. J.*, **18**, 6885–6894.
223. Guo, C., Song, J., and Gong, L.-Z. (2013) *Org. Lett.*, **15**, 2676–2679.
224. Shi, F., Zhu, R.-Y., Dai, W., Wang, C.-S., and Tu, S.-J. (2014) *Chem. Eur. J.*, **20**, 2597–2604.
225. Luo, W., Lin, Y., Yang, D., and He, L. (2014) *Tetrahedron: Asymmetry*, **25**, 787–791.
226. Shi, F., Luo, S.-W., Tao, Z.-L., He, L., Yu, J., Tu, S.-J., and Gong, L.-Z. (2011) *Org. Lett.*, **13**, 4680–4683.
227. Shi, F., Tao, Z.-L., Yu, J., and Tu, S.-J. (2011) *Tetrahedron: Asymmetry*, **22**, 2056–2064.
228. Shi, F., Zhu, R.-Y., Liang, X., and Tu, S.-J. (2013) *Adv. Synth. Catal.*, **355**, 2447–2458.
229. Shi, F., Xing, G.-J., Tan, W., Zhu, R.-Y., and Tu, S. (2013) *Org. Biomol. Chem.*, **11**, 1482–1489.
230. Yu, J., He, L., Chen, X.-H., Song, J., Chen, W.-J., and Gong, L.-Z. (2009) *Org. Lett.*, **11**, 4946–4949.
231. Wang, C.-S., Zhu, R.-Y., Zheng, J., Shi, F., and Tu, S.-J. (2014) *J. Org. Chem.*, **80**, 512–520.
232. Liu, W.-J., Chen, X.-H., and Gong, L.-Z. (2008) *Org. Lett.*, **10**, 5357–5360.

233. Zhu, R.-Y., Wang, C.-S., Jiang, F., Shi, F., and Tu, S.-J. (2014) *Tetrahedron: Asymmetry*, **25**, 617–624.
234. Li, N., Song, J., Tu, X.-F., Liu, B., Chen, X.-H., and Gong, L.-Z. (2010) *Org. Biomol. Chem.*, **8**, 2016–2019.
235. Zamfir, A. and Tsogoeva, S.B. (2011) *Synthesis*, **2011**, 1988–1992.
236. Serdyuk, O.V., Zamfir, A., Hampel, F., and Tsogoeva, S.B. (2012) *Adv. Synth. Catal.*, **354**, 3115–3121.
237. Rueping, M., Maji, M.S., Küçük, H.B., and Atodiresei, I. (2012) *Angew. Chem. Int. Ed.*, **51**, 12864–12868.
238. Hashimoto, T., Omote, M., and Maruoka, K. (2011) *Angew. Chem. Int. Ed.*, **50**, 3489–3492.
239. Zhu, R.-Y., Wang, C.-S., Zheng, J., Shi, F., and Tu, S.-J. (2014) *J. Org. Chem.*, **79**, 9305–9312.
240. Hong, L., Kai, M., Wu, C., Sun, W., Zhu, G., Li, G., Yao, X., and Wang, R. (2013) *Chem. Commun.*, **49**, 6713–6715.
241. Müller, S. and List, B. (2009) *Angew. Chem. Int. Ed.*, **48**, 9975–9978.
242. Müller, S. and List, B. (2010) *Synthesis*, (13), 2171–2178.
243. Das, A., Volla, C.M.R., Atodiresei, I., Bettray, W., and Rueping, M. (2013) *Angew. Chem. Int. Ed.*, **52**, 8008–8011.
244. Rueping, M. and Antonchick, A.P. (2008) *Angew. Chem. Int. Ed.*, **47**, 10090–10093.
245. Lee, S. and Kim, S. (2009) *Tetrahedron Lett.*, **50**, 3345–3348.

3
Reactions of Generated Imine Intermediates

3.1
Enamines

3.1.1
Nucleophilic Addition Reactions

Although imines dominate the landscape of chiral Brønsted acid catalysis, they also suffer with stability issues. Enamines offer a convenient, stable, and easy-to-handle source of imines, but surprisingly their usage has been rather limited. One possible reason for this is that they are generally employed as nucleophilic rather than electrophilic components. In 2008, Tsogoeva realized the potential of enamides to behave as both a nucleophile and an electrophile and thus undergo a self-coupling reaction for the formation of quaternary carbon centers. By treatment of enamides **1a** with 10 mol% **PA 8**, it was found that a facile self-coupling was possible, leading to the products **1b** with high levels of selectivity (Scheme 3.1) [1].

Not only did the products **1b** possess a quaternary center, which is generally considered a difficult feat to control, but they also resemble useful chiral building blocks, which were shown to be suitable for further transformations.

The mechanism of self-coupling is illustrated in Scheme 3.2. Firstly, one molecule of enamine must rearrange itself to its corresponding imine tautomer. Once this has occurred, it can then be activated by the catalyst through hydrogen bonding while also promoting the attack of the non-tautomerized enamine. This will initially yield an iminium intermediate, which can release the product by rearrangement and proton transfer back to the catalyst. A related Mannich reaction using enecarbamates was also disclosed by Terada in the same year [2].

The Friedel–Crafts reaction is another protocol that has seen intensive developments with respect to imine coupling partners, but with enamines, only two such reports exist. In the same year, the groups of Terada and Zhou both disclosed their respective results on the use of enamines toward this reaction. Terada showed that Boc-protected enecarbamates **2a** would undergo coupling with indoles **2b** in the presence of **PA 16** to give the corresponding products **2c** with excellent levels of

Asymmetric Brønsted Acid Catalysis, First Edition. Magnus Rueping, Dixit Parmar and Erli Sugiono.

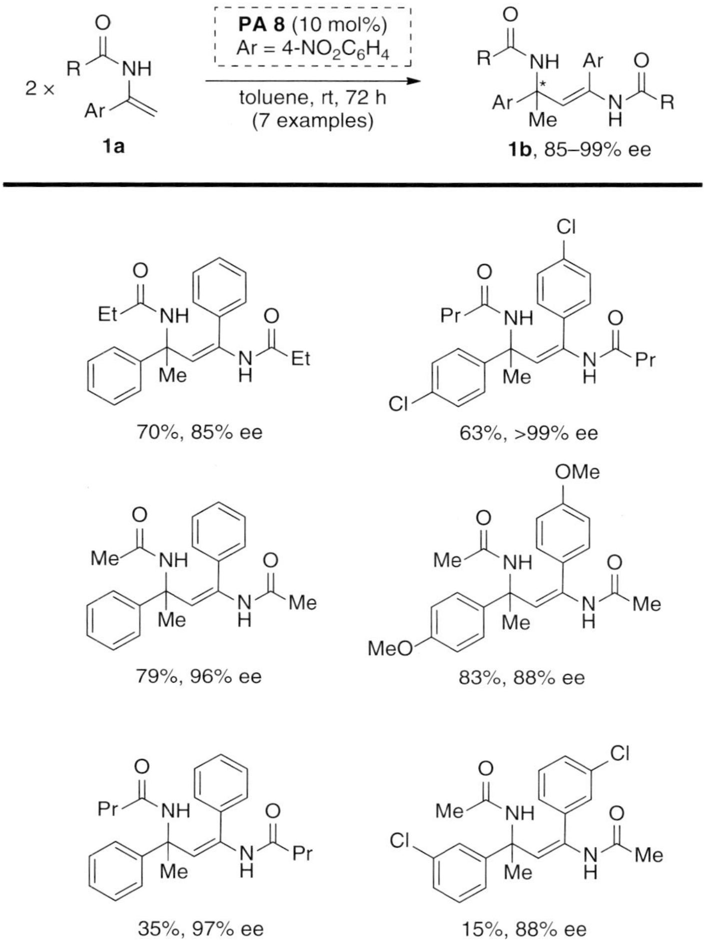

Scheme 3.1 Self-coupling of enamides by Tsogoeva [1].

enantioselectivity (Scheme 3.3) [3]. Since enecarbamates **2a** were shown to fully isomerize to the corresponding imines during distillation, it was concluded that this was the active component during the reaction.

Similarly, Zhou showed that he could employ α-aryl enamides **2d** as suitable partners to indoles **2e** to construct a quaternary center at the 3-position of indoles **2f** also using a phosphoric acid catalyst (Scheme 3.4) [4]. Zhou performed a selection of mechanistic experiments, which showed that both N–H-free enamides and N–H-free indoles were required for reactivity; when either was protected, no reaction occurred. In transformation-related work, Bandini has shown that all enamides can be employed as suitable partners for indoles [5], and You has shown generated enamines from pyridines to be suitable for pyrrole nucleophiles [6].

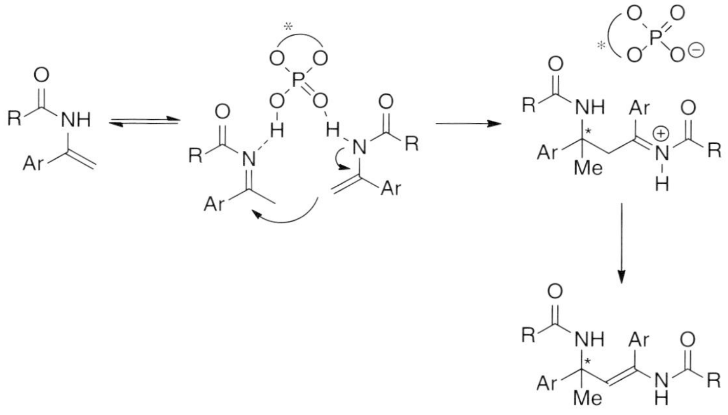

Scheme 3.2 Mechanism of self-coupling [1].

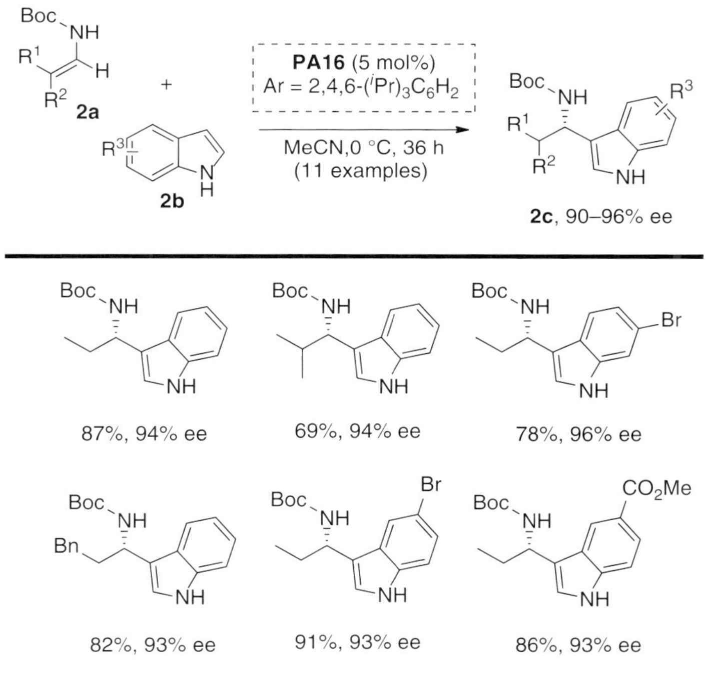

Scheme 3.3 Friedel–Crafts reaction of enamides by Terada [3].

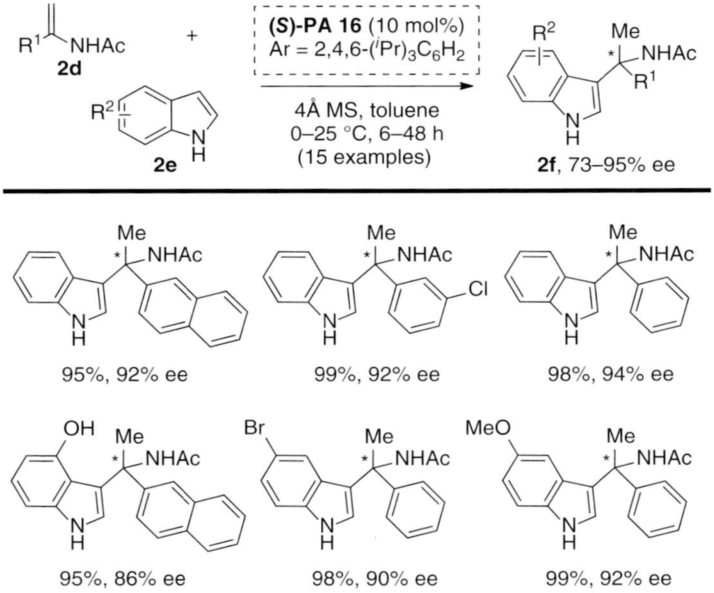

Scheme 3.4 Friedel–Crafts reaction of enamides by Zhou [4].

3.1.2
Transfer Hydrogenation

Asymmetric transfer hydrogenation is one of the most efficient transformations to access enantioenriched products, and while imines serve as proficient substrates, enamines can offer several advantages as substrates to access very similar products. The only report in this area has been shown by the group of Antilla, who reported on the transfer hydrogenation of enamides **3a** by the use of Hantzsch ester and a combination of **(*S*)-PA 1** and achiral acid AcOH (Scheme 3.5) [7].

A scope that included only aromatic enamides was explored, and generally excellent levels of enantioselectivity were obtained; however, alkyl enamides failed to react under these conditions. Although the use of AcOH was not necessary for the reaction to proceed, far greater yields of the products were obtained when utilized and it was proved that its role is to aid transformation of the enamide to the corresponding imine. The ability of acid catalysts to catalyze a series of sequential steps is also a useful property, and this has led groups to perform cascade reactions. In 2008, Rueping reported one such cascade involving a transfer hydrogenation step of the coupling of enamines **4a** and α,β-unsaturated ketones **4b** (Scheme 3.6) [8].

The overall efficiency of the cascade and the levels of enantioselectivity of the tetrahydropyridine products **4c** are remarkable, considering the simplicity of the starting materials.

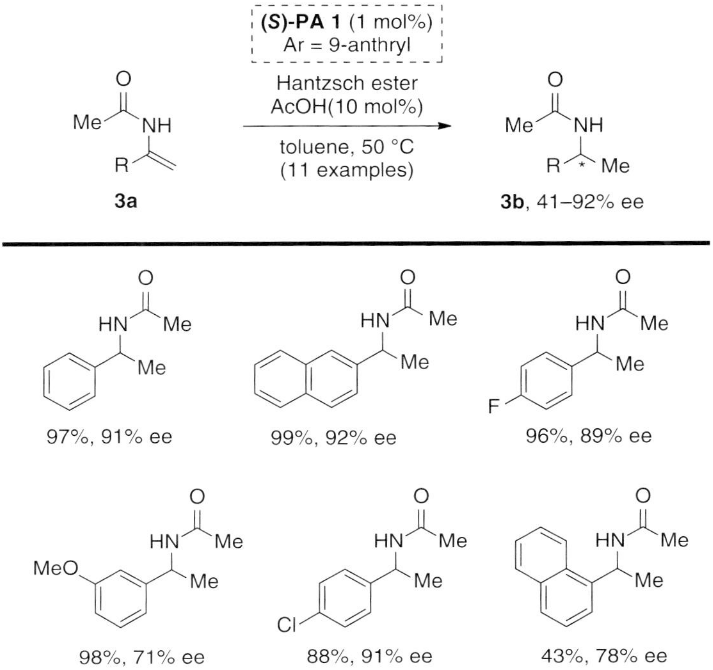

Scheme 3.5 Transfer hydrogenation of enamides by Antilla [7].

The mechanism of the cascade can be observed in Scheme 3.7. The process starts with the conjugate addition of the enamine to the α,β-unsaturated ketone, which may well be catalyzed by the chiral acid catalyst. This leads to an intermediate, which can undergo the loss of H_2O by condensation of the amine and ketone units. The cyclic imine formed can then be activated by the phosphoric acid, which also promotes the addition of a hydride from Hantzsch ester in an asymmetric manner to yield the observed products.

3.2 Indoles Containing Leaving Groups

Indoles containing leaving groups may at first glance sound like an unusual category of substrates, especially for activation by Brønsted acids, but in fact, a considerable level of attention has been dedicated by research groups to this class of compounds. A common feature that most substrates share in this category is the presence of a good leaving group under acidic conditions (e.g., OH) at the 3-position. By elimination of this small molecule, a highly electrophilic iminium-like species is formed, which can be intercepted by a host of nucleophiles in an asymmetric manner. This general mode of reactivity can be assumed to be occurring in this category unless otherwise stated.

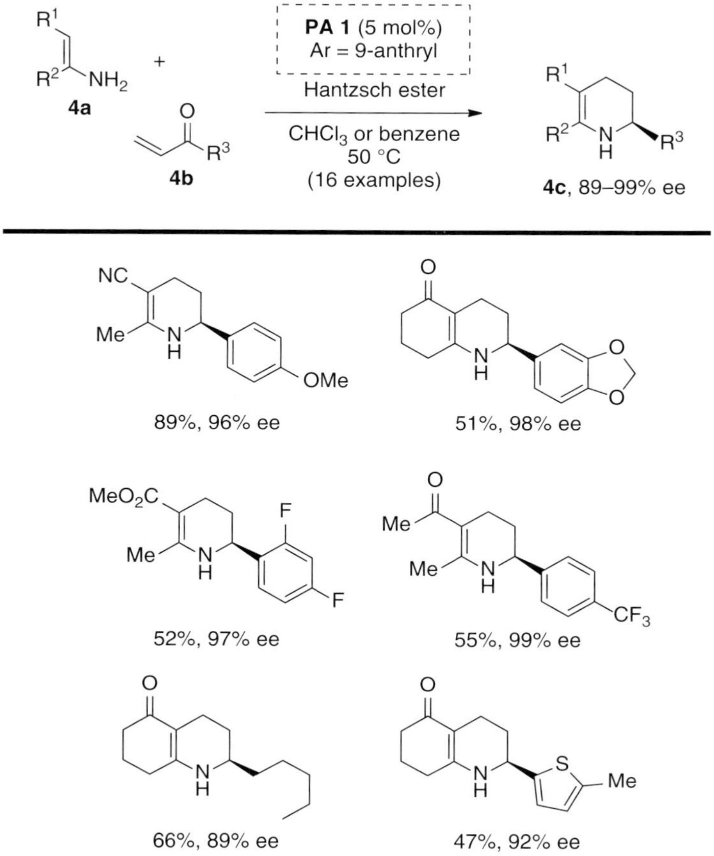

Scheme 3.6 Cascade reaction involving a transfer hydrogenation by Rueping [8].

3.2.1 Nucleophilic Addition Reactions

One of the seminal reports within this unique category was reported by Gong in 2009, who showed the coupling of aromatic enamides **5a** with 3-hydroxy indoles **5b** in the presence of phosphoric acid catalyst **[H$_8$]-PA 12** to yield propanone substituted indoles **5c** in high enantioselectivities (Scheme 3.8) [9].

During the catalyst optimizations, it was found that sterically hindered catalysts generally gave lower levels of selectivity, while a simple phenyl-substituted octahydro-catalyst gave the optimal results. This feature of chiral phosphoric acid catalysts occurs more commonly than thought and suggests that more complex factors are at play in the transition state such as solvation effects.

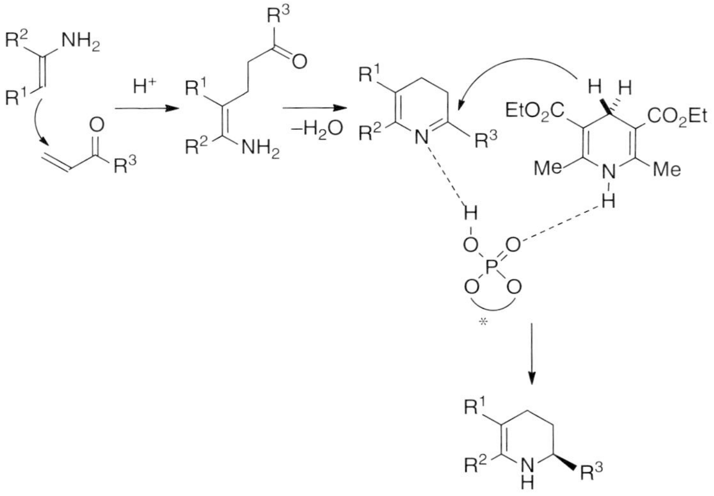

Scheme 3.7 Mechanism of the cascade reaction [8].

With regard to the mechanism, the authors propose that the reaction is controlled by a chiral phosphate counterion, which can be seen in Scheme 3.9. Elimination of H_2O is the first step in the process, which results in the formation of a cationic intermediate containing a bound phosphate anion. The intermediate is proposed to exist either as an iminium ion or as a carbocation. Attack of the enamine on this electrophilic species results in a new iminium species, which can be hydrolyzed by H_2O to yield the products. Gong has also employed this methodology as a key step in his total synthesis of (+)-folicanthine [10].

Closely related to enamine nucleophiles are carbonyl substrates, which are inherently nucleophilic in their enol form. In 2012, Guo and Peng developed the coupling of 3-hydroxy indoles **6a** with cyclohexanone **6b** to yield highly complex 3-indolyloxindoles **6c** with high levels of enantioselectivity (Scheme 3.10) [11]. The reaction explores a broad scope and was even shown to function well with a series of cyclic and acyclic substrates. Mechanistic work revealed that the presence of an N–H-free indole is crucial to achieving enantiocontrol, suggesting that the catalyst (**PA 25**) may be involved in coordination with this proton.

Recently, Shi has explored the coupling of 3-hydroxy indoles **6d** with 2-hydroxy styrenes **6e** to yield the corresponding allylation product **6f** with excellent enantioselectivities (Scheme 3.11) [12]. Detailed experiments also showed that in this work, both N–H-free indole and the hydroxyl group of the nucleophile were essential to reactivity and enantioselectivity. Shi has also disclosed various extensions on the use of 3-hydroxy indoles with different nucleophiles [13–15].

The Pinacol rearrangement is a useful transformation for the synthesis of 1,2-substituted ketones from the corresponding 1,2-diols via a dehydration process. The major difficultly encountered is the unpredictability of regioselectivity as usually both –OH groups of the diol can potentially form the ketone moiety. In 2010,

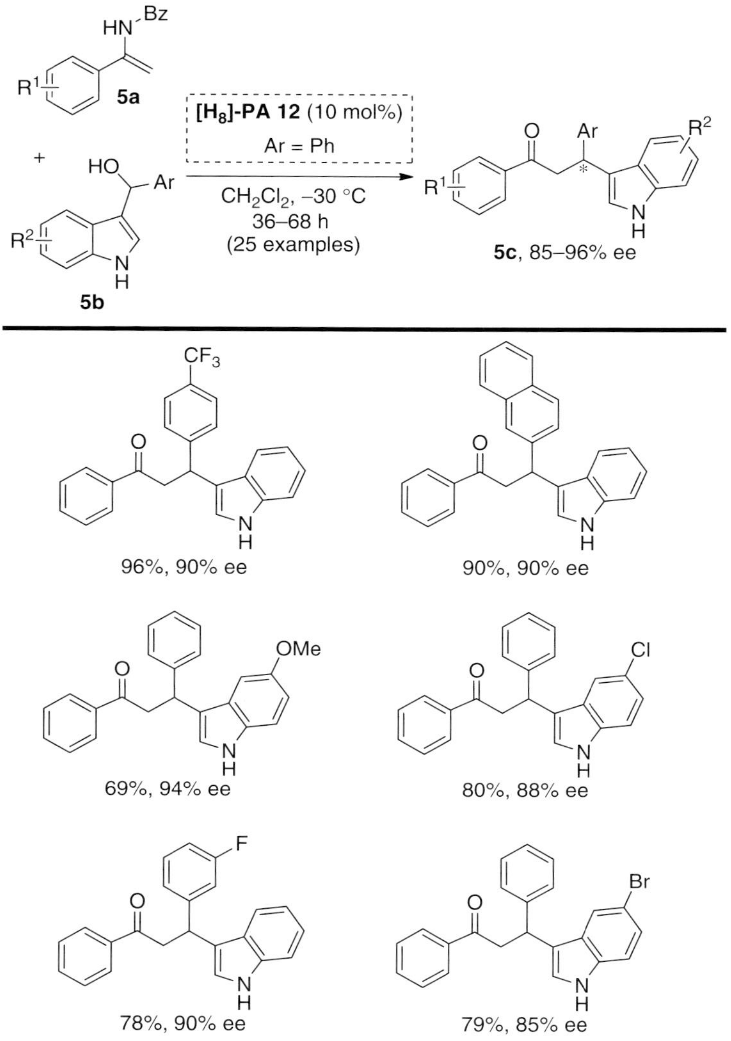

Scheme 3.8 Coupling of enamides with 3-hydroxy indoles by Gong [9].

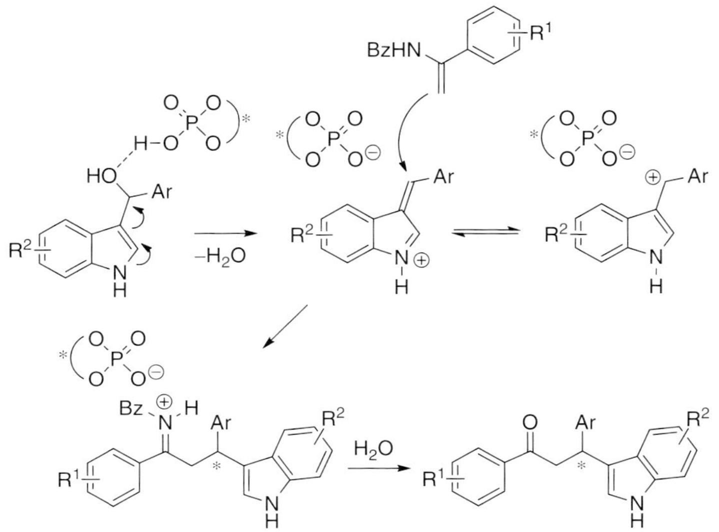

Scheme 3.9 Mechanism of enamide and 3-hydroxy indole couplings [9].

Antilla devised a clever strategy of employing 3-hydroxy indoles **7a**, which can preferentially dehydrate at the indolyl-alcohol center to undergo a Pinacol rearrangement to yield indolyl ketone **7b** (Scheme 3.12) [16].

A broad scope was explored using only 2.5 mol% of catalyst **[H$_8$]-PA 14,** which generally delivered very high levels of stereocontrol in the products. In all cases, the indole nitrogen remained protected and so the stereoselectivity is believed to arise from a chiral phosphate anion. The proposed mechanism is given in Scheme 3.13.

It is believed that coordination occurs between both hydroxyl groups and the catalyst, which allows the indole's nitrogen to push electron density within its ring system to eliminate H_2O. Upon elimination, a cationic iminium ion is formed and the remaining tertiary hydroxyl group can cause a 1,2-migration onto the newly formed electrophilic center to yield the desired products. Although a hydroxyl group is actually preferred by most research groups' methodology, the use of *N*-sulfonamides as a motif suitable for displacement by thiols has been reported by Tian [17].

A rather unique example both from a substrate and from a transformation point of view was presented by Terada in 2011, for which he described the stereoselective addition of azlactones **8b** to 3-vinyl indoles **8a** (Scheme 3.14) [18].

The protocol reported is a two-step procedure whereby the addition of **8b** to the vinyl group occurs first followed by hydrolysis of this intermediate to yield the tryptophan derivatives **8c** with good to high enantioselectivities. The authors propose that the addition of azlactone may occur concomitantly with protonation; however, the generation of a discreet iminium-like species cannot be ruled out.

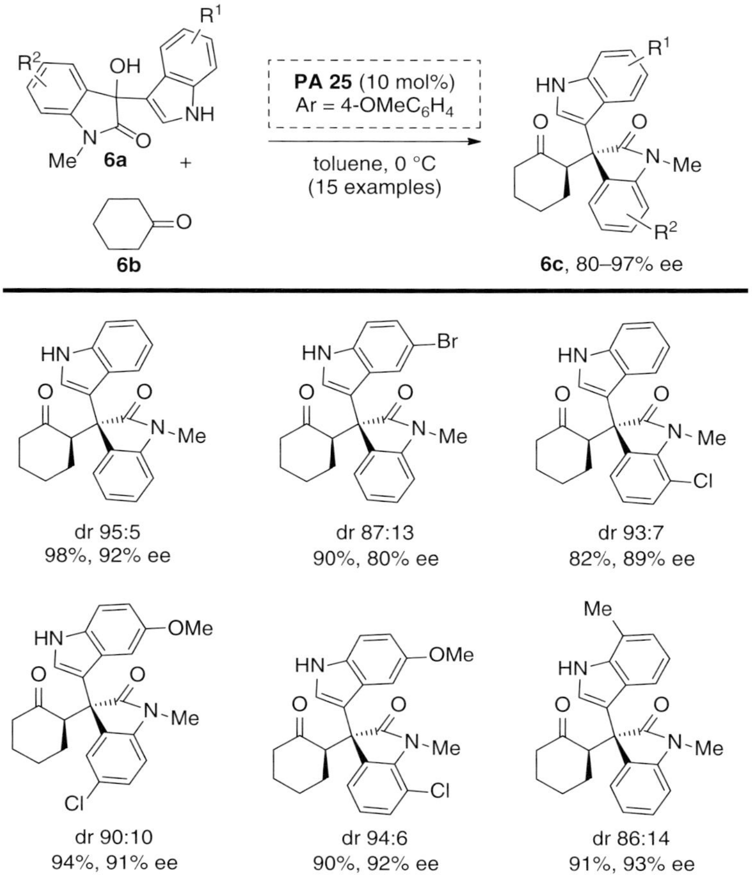

Scheme 3.10 Coupling of ketones by Guo and Peng [11].

3.2.2
Friedel–Crafts Reactions

The Friedel–Crafts reaction is a well-studied procedure for the employment of chiral Brønsted acids, and the activation of electrophilic 3-substituted indoles has proved to provide moderate success for research groups. In 2010, You showed that 3-indolyl amines **9a** could serve as efficient partners for methyl-protected indole **9b** in performing a Friedel–Crafts reaction to access bisindole products **9c** with modest selectivities (Scheme 3.15) [19]. More recently, Shi has shown that N–H-free indoles work well with 3-indolylmethanols [20].

More recently, Jiang and Zhang have developed an imidodiphosphoric acid catalyst **NPA 5,** which can be used for the synthesis of 1-indolyl pyrroles **9f** with high enantioselectivities (Scheme 3.16) [21]. They opted to utilize 3-protected

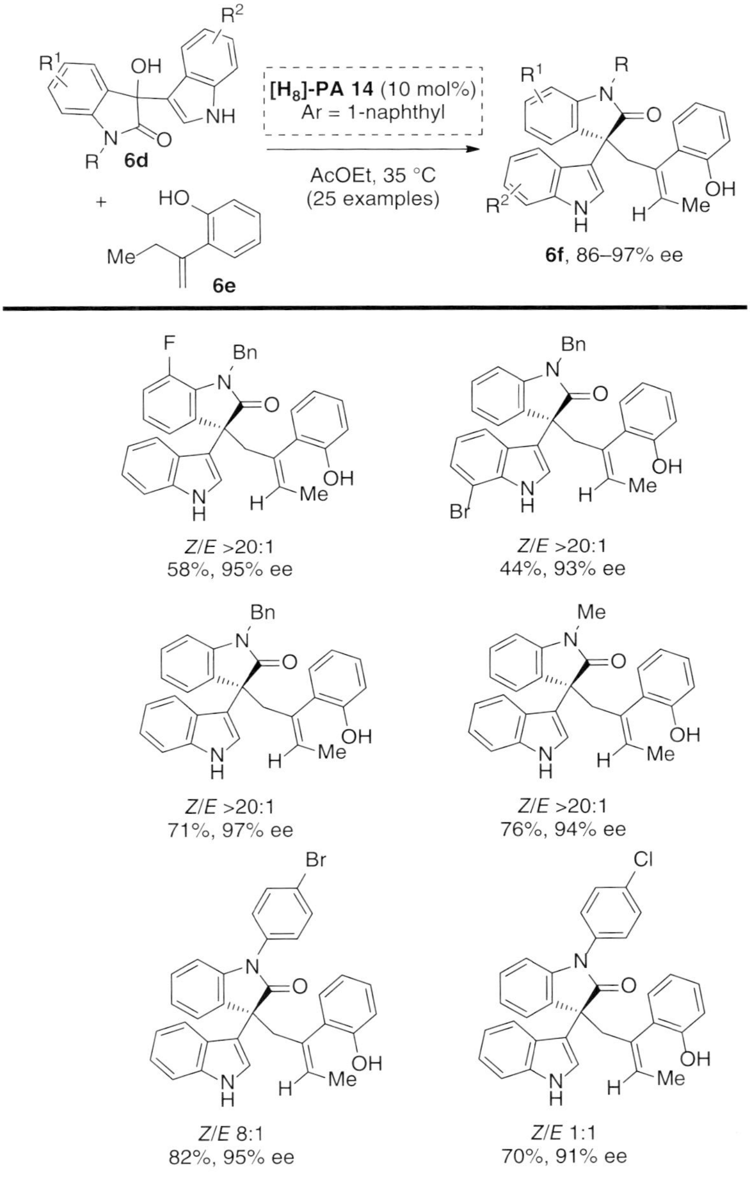

Scheme 3.11 Coupling of 2-hydroxy styrenes with 3-hydroxy indoles by Shi [12].

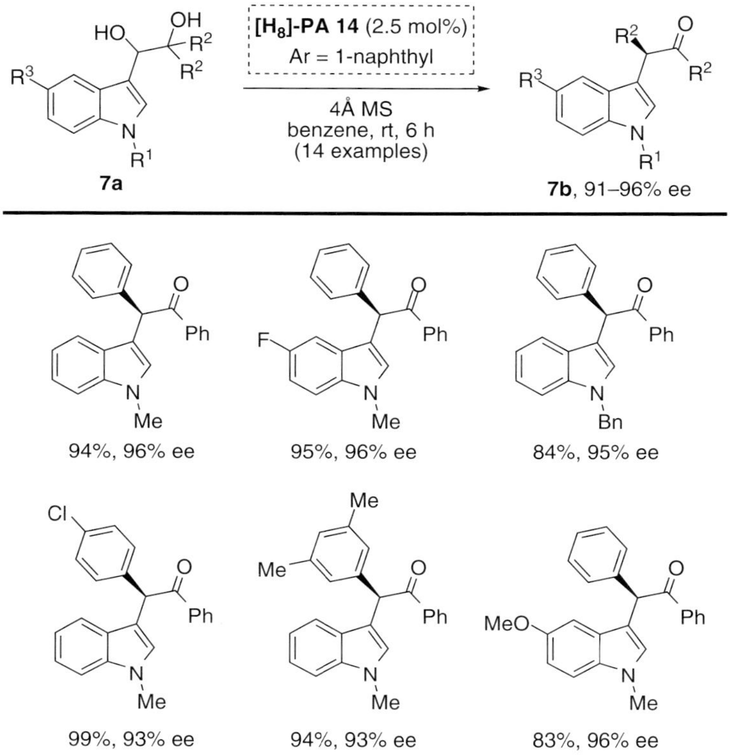

Scheme 3.12 Pinacol rearrangement of 3-hydroxy indoles by Antilla [16].

hydroxy-indoles as precursors for the generation of electrophilic iminium species to undergo couplings with pyrroles.

The key part of the mechanism of this reaction is very similar to the previously described mechanisms for 3-indolyl alcohols and is portrayed in Scheme 3.17. A catalyst-promoted elimination step initiates the process, which leads to a highly electrophilic iminium ion containing a bound chiral phosphate anion. This is intercepted by the indole component, which results in the observed product. Recently, leaving groups positioned at the 2-position of indoles have also been shown to be suitable for participating in Friedel–Crafts reactions [22].

An earlier but more elaborate report was by You in 2009, who demonstrated that aromatic aldehyde **10a** and N–H-free indoles **10b** could participate in a double Friedel–Crafts reaction to form enantioenriched fluorenes **10c** (Scheme 3.18) [23, 24].

The mechanism of this cascade is illustrated in Scheme 3.19. The first Friedel–Crafts reaction between the aldehyde and the indole yields a 3-hydroxy indole motif. As we have seen within this section, in the presence of the acidic catalyst, this undergoes a dehydration to reveal an electrophilic iminium ion. This can be attacked by the pendant highly electron-rich aromatic group to yield the

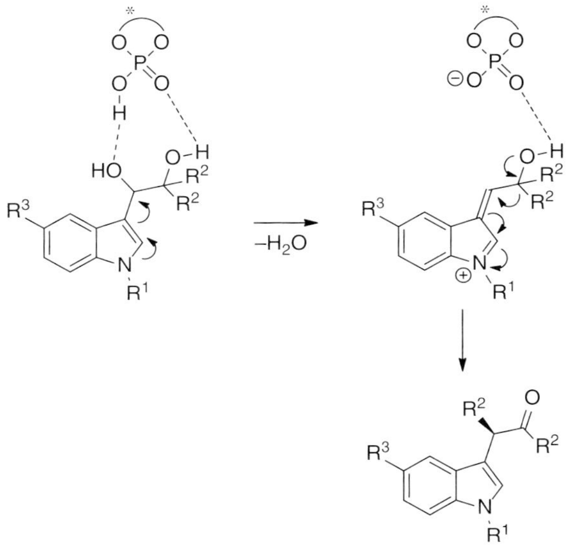

Scheme 3.13 Mechanism of pinacol rearrangement [16].

products. Mechanistic experiments supported this order of events and showed that N–H-free indole was crucial to achieving high levels of enantioselectivity.

3.3
N-Acetals and Aminals

N-Acetals/aminals share a significance with the structural motifs discussed in the previous subsection (3-substituted indoles), notably containing the opportunity to eliminate a small molecule to generate an iminium-like species. These iminium species are highly electrophilic and have been shown by research groups to participate in a selection of interesting and useful asymmetric methodologies.

3.3.1
Nucleophilic Addition Reactions

Although it is well accepted that nitrogen-containing acetals under acidic conditions will generate imines or iminium-type species, their reactivity profiles should not be considered equal, and, thus, the use of *N*-acetals may allow unique transformations to be developed. In 2009, this was discovered to be the case for Terada, who found that activation of hemiaminal ethers **11a** in the presence of **PA 16** or **27** allowed a stereoselective addition of enecarbamates **11b** to occur, yielding the diamine products **11d** after reduction (Scheme 3.20) [25].

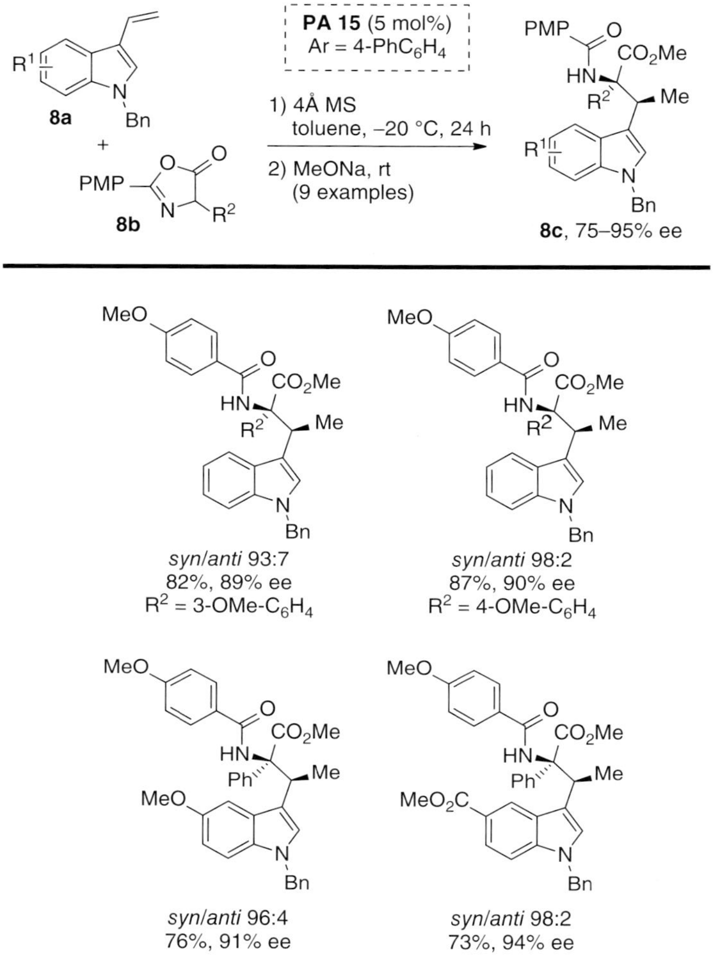

Scheme 3.14 Additions to 3-vinyl indoles by Terada [18].

The reaction generally proceeds well to give the desired products with good levels of enantioselectivity. It was found that the methanol released from **11a** during activation plays a crucial role in the reaction, and the proposed mechanism is shown in Scheme 3.21.

The reaction commences by the loss of MeOH from the hemiaminal ether, which may be promoted by the catalyst. Upon loss, an imine is formed, and this can be activated toward attack by the enecarbamate. This generates an iminium intermediate, which has the potential to undesirably go on and react further with another equivalent of enecarbamate. Preventing this from occurring is the MeOH, which now comes back and traps this intermediate as a hemiaminal ether. After the completion of the reaction, a simple reduction step using K-selectride

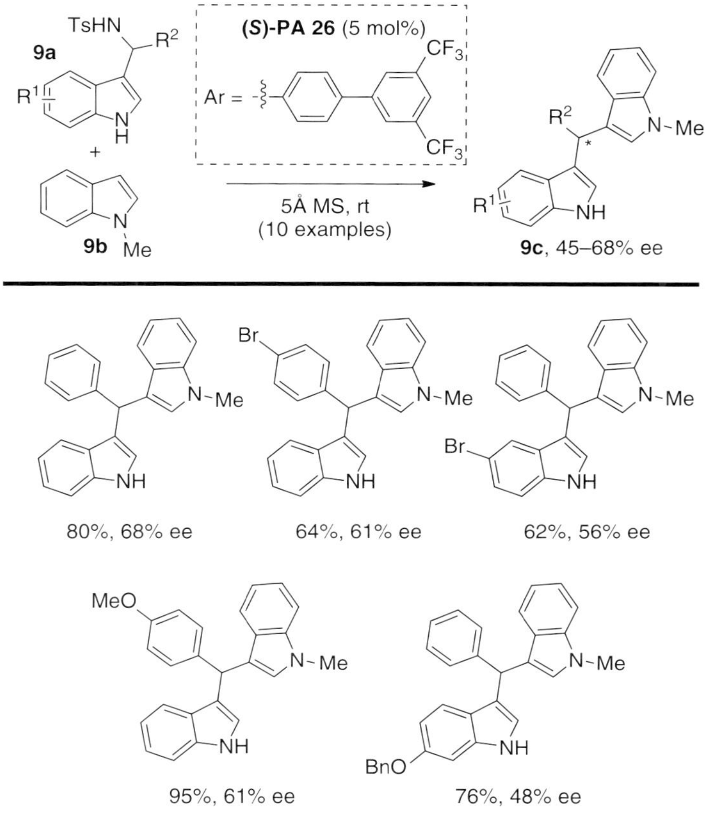

Scheme 3.15 Synthesis of triarylmethanes by You [19].

yields the desired products. The concept of an eliminated group playing a dual role has also been exploited by Terada, who reported the Aza-Petasis–Ferrier rearrangement of hemiaminal ethers **12a** (Scheme 3.22) [26, 27].

The mechanism of this reaction is illustrated in Scheme 3.23. Treatment of the hemiaminal ether with the catalyst initially leads to the elimination of vinyl enol ether and an imine. This can be activated by the catalyst to undergo coupling to yield β-amino aldehydes. A subsequent reduction by $NaBH_4$ yields the desired products. Mechanistic experiments appear to reveal that the origin of stereoselectivity arises from the starting hemiaminal ethers. Support for this notion was found when the use of an enantiopure hemiaminal ether resulted in a higher enantiopurity of the desired product. Recently, Shi has reported the addition of *o*-hydroxystyrenes to hydroxylactams [28].

In 2013, Maruoka developed the asymmetric addition of β-keto esters **13b** to aminals **13a** using a highly sterically hindered catalyst **(*S*)-PA 28** in 10 mol% (Scheme 3.24) [29]. The reaction can formally be thought of as a Mannich reaction and the products **13c** were obtained with very high enantioselectivities.

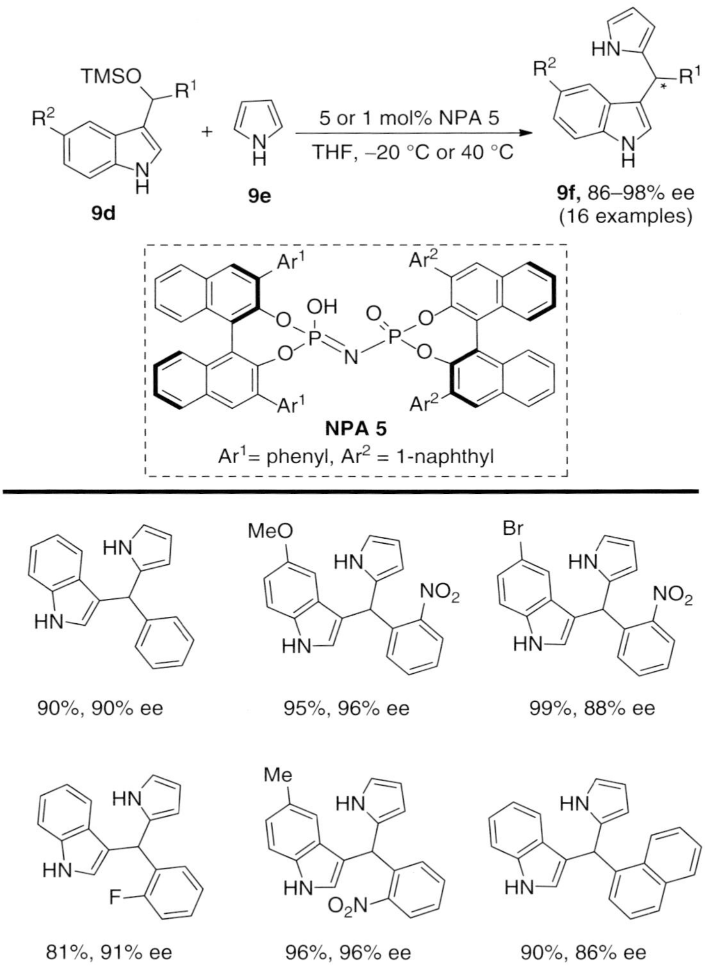

Scheme 3.16 Synthesis of triarylmethanes by Jiang/Zhang [21].

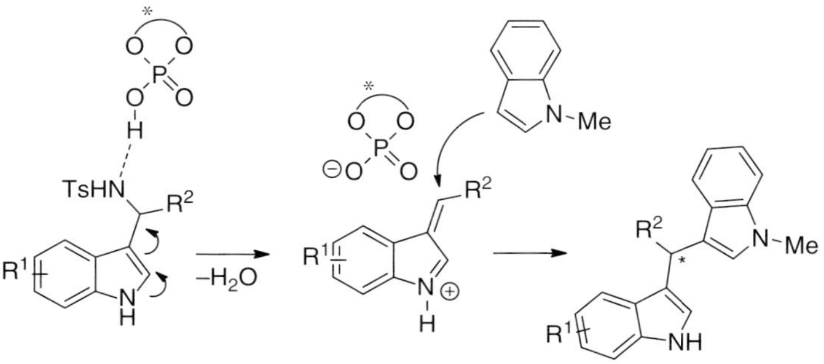

Scheme 3.17 Mechanism of Friedel–Crafts reaction [19].

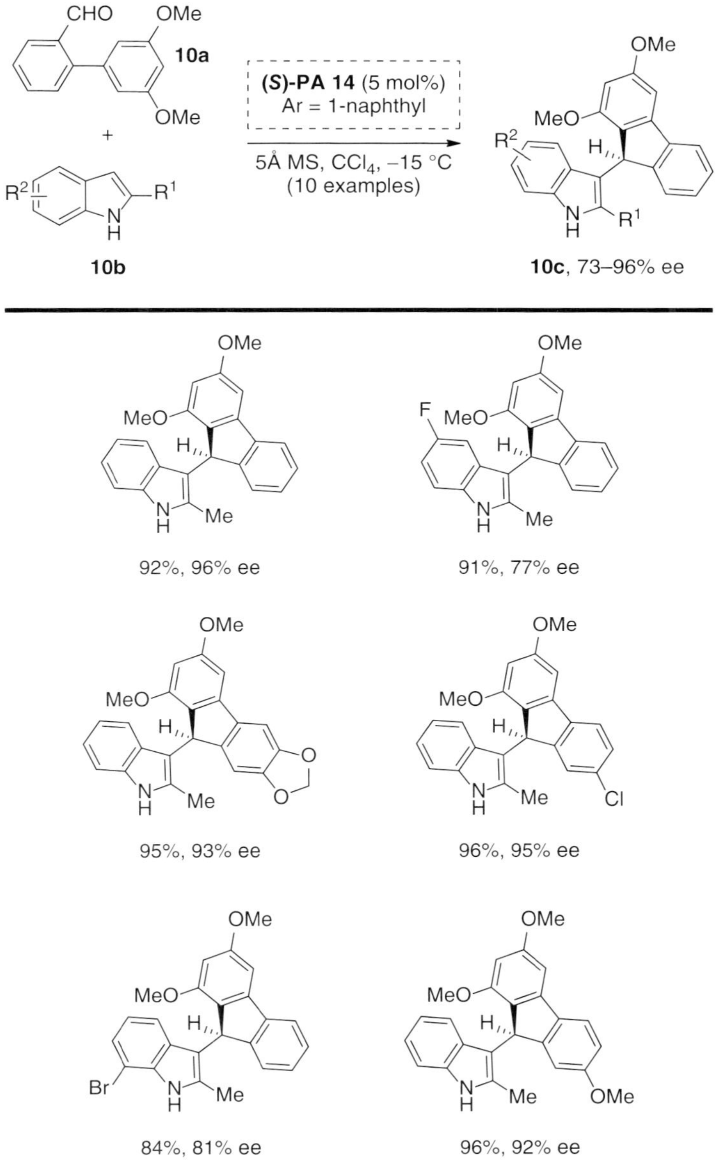

Scheme 3.18 Double Friedel–Crafts reaction by You [23].

In general, the reaction works well for a series of alkyl and aromatic substitution on the alkyne and with slightly modified conditions can also be extended to acyclic β-keto esters. The products themselves also serve as useful synthetic building blocks for further transformations.

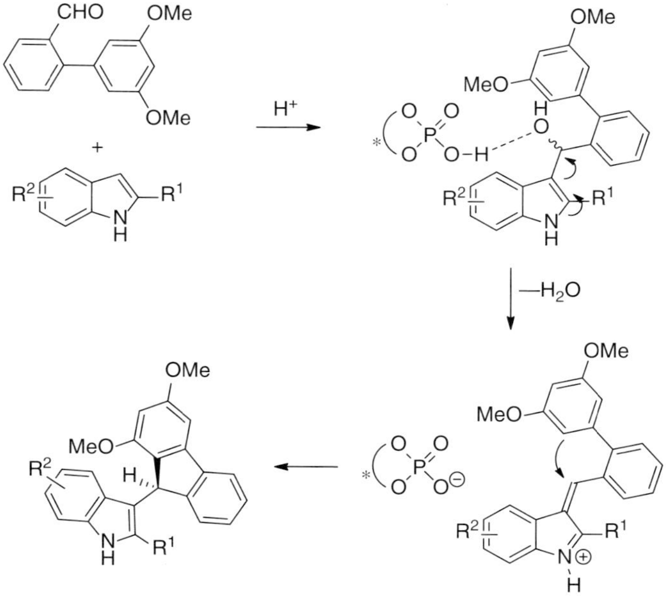

Scheme 3.19 Mechanism of double Friedel–Crafts reaction [23].

3.3.2 Friedel–Crafts Reactions

The Friedel–Crafts reaction can be considered as one of the most common testing grounds for the development of new electrophilic species toward undergoing asymmetric reactions under Brønsted acid catalysis. In 2010, Rueping designed one such reactions of indole **14a** with γ-hydroxylactams **14b** in the presence of 5 mol% **[H$_8$]-NPA 6** (Scheme 3.25) [30].

In 2012, Lete also demonstrated this reaction in a more complex setting to access isoindoloisoquinolines **14f** (Scheme 3.26) [31, 32]. Coupling of various indoles **14e** with γ-hydroxylactams **14d** gave the desired products in modest yields and selectivities. The mechanisms for both processes share similar traits and a representative illustration is shown in Scheme 3.27. It is thought that under the acidic conditions, loss of H_2O occurs to reveal an iminium ion containing a chiral phosphate anion. This is then intercepted by a suitable nucleophile to yield the addition products. Several research groups have also explored this concept with chiral phosphoric acids [33–35].

The activation of cyclic acetals by Brønsted acids was also shown by You in 2011. He reported on the coupling between spiroindoline-2-ones **15a** and various indoles **15b** to give the Friedel–Crafts adducts **15c** in the presence of **(*S*)-PA 16** (Scheme 3.28) [36].

The scope of the reaction was quite broad and could also be extended to pyrroles and benzene coupling partners. In general, the products were obtained

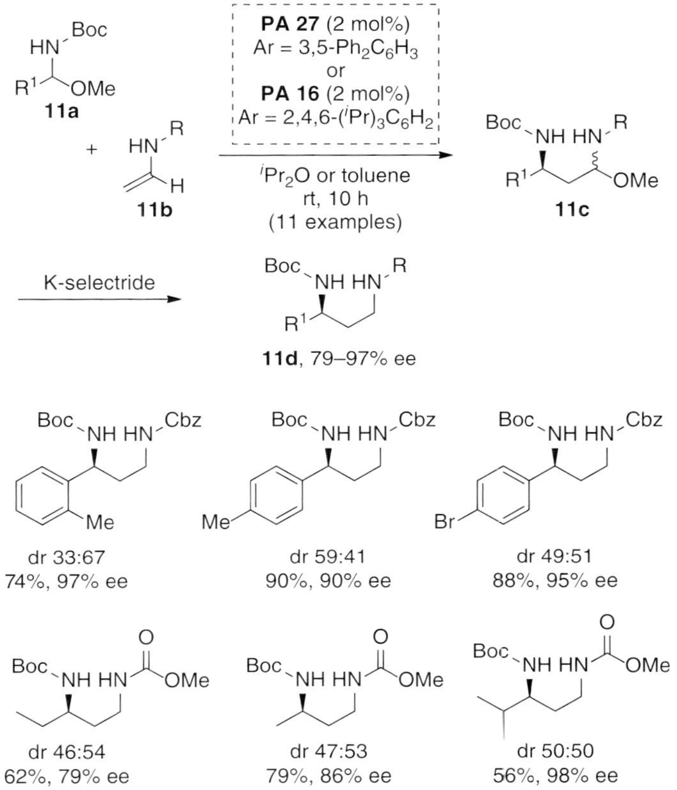

Scheme 3.20 Coupling of enecarbamates with hemiaminal ethers by Terada [25].

with excellent enantioselectivities and were shown to be useful for further transformations. The mechanism of the reaction differs slightly from the previously described reports as in this case, a molecule is not formally eliminated from the starting material. The proposed reaction sequence is portrayed in Scheme 3.29. The starting material must tautomerize itself from the hemiaminal ether form to a hydroxyimine. This can then be activated by the catalyst and allows for the indole to attack the electrophilic carbon to yield the desired products.

3.3.3
Transfer Hydrogenation

The asymmetric transfer hydrogenation of imines represents an important transformation to access chiral amines, and so, the opportunity to use imine precursors is an appealing extension. The seminal report for the use of *N,O*-acetals came from Zhou in 2012, who showed that γ-hydroxylactams **16a** could undergo hydrogenation with a Hantzsch ester in the presence of VAPOL-derived catalyst **PA 6** (Scheme 3.30) [37].

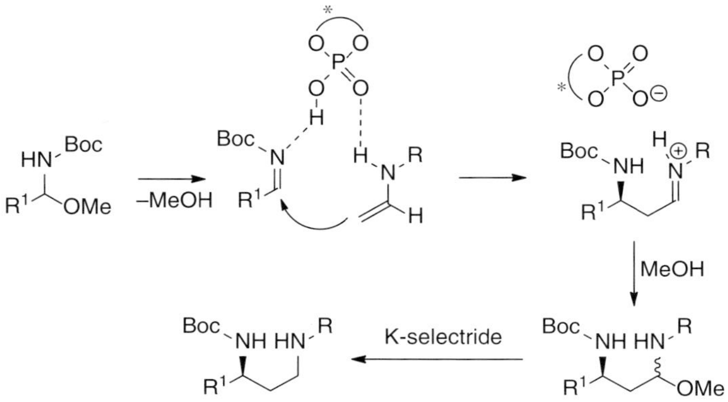

Scheme 3.21 Mechanism of enecarbamate couplings with hemiaminal ethers [25].

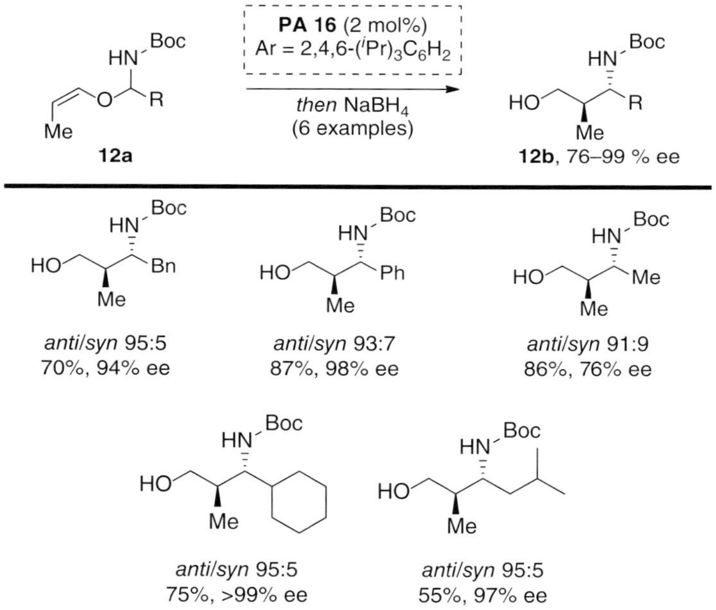

Scheme 3.22 Aza-Petasis–Ferrier rearrangement by Terada [26].

The reaction shows a broad scope and delivers the products **16b** in modest yields and with good enantioselectivities. The presence of the N–H on the lactam was shown to be crucial, as a methyl-protected lactam failed to yield any desired product. Similar hydrogenation reactions have also been reported by You [38] and Jia [39].

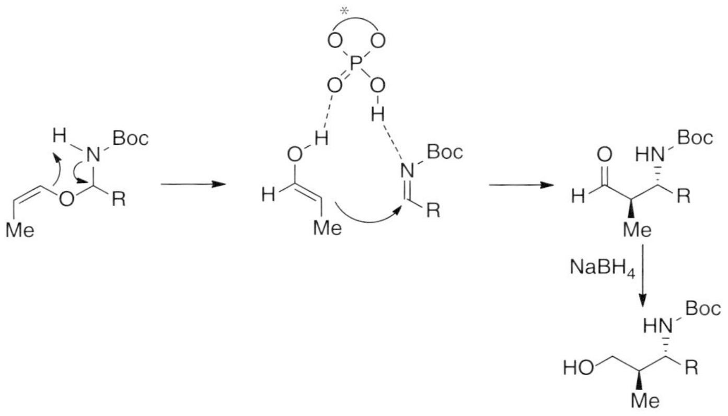

Scheme 3.23 Mechanism of aza-Petasis–Ferrier rearrangement [26].

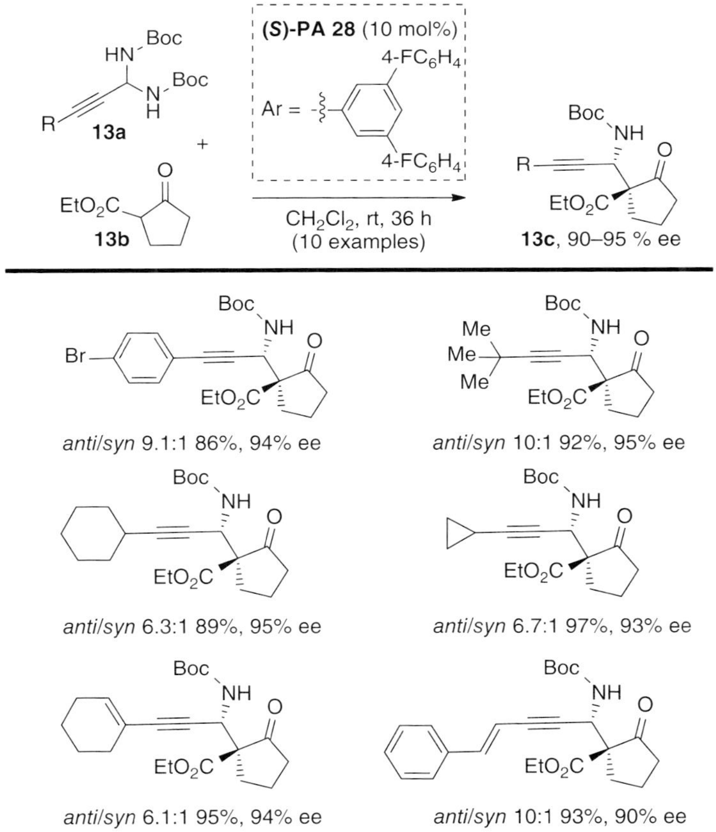

Scheme 3.24 Coupling of aminals with β-keto esters by Maruoka [29].

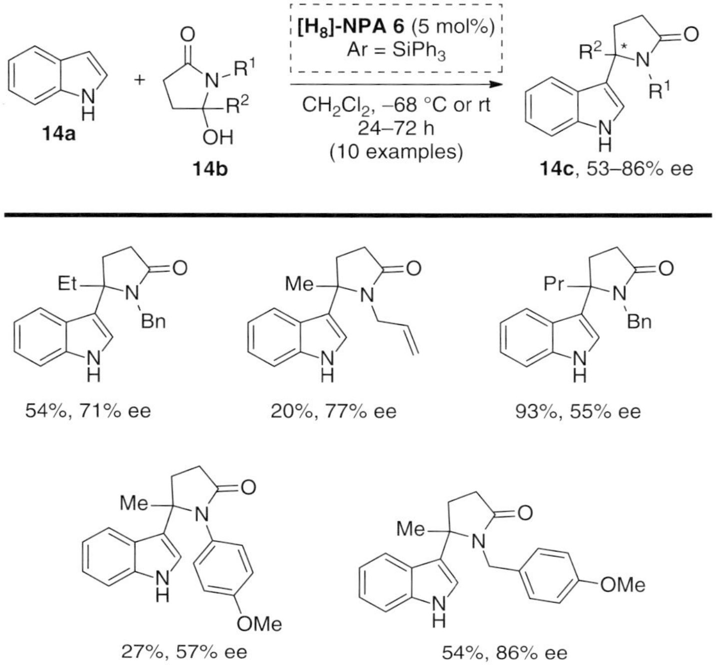

Scheme 3.25 Addition to chiral *N*-acyliminium ions by Rueping [30].

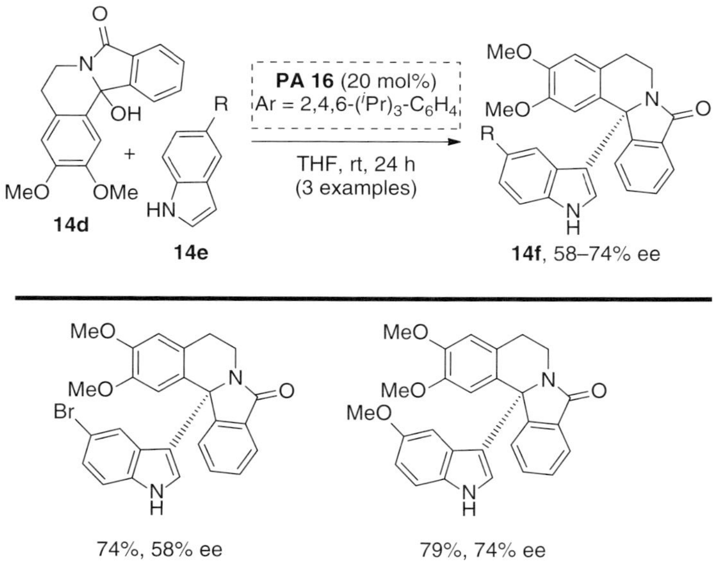

Scheme 3.26 Addition to chiral *N*-acyliminium ions by Lete [31, 32].

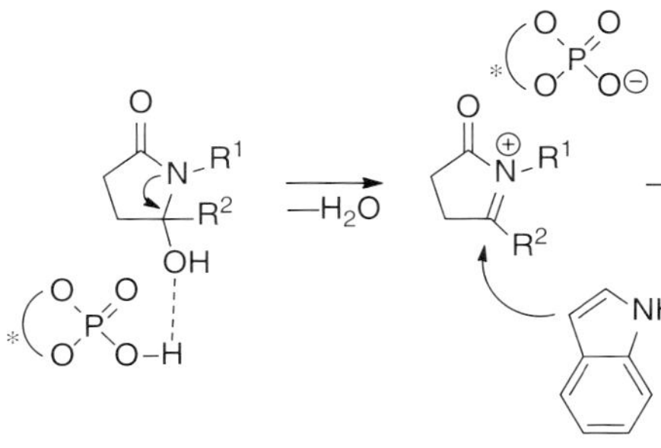

Scheme 3.27 Mechanism of addition to chiral *N*-acyliminium ions [30].

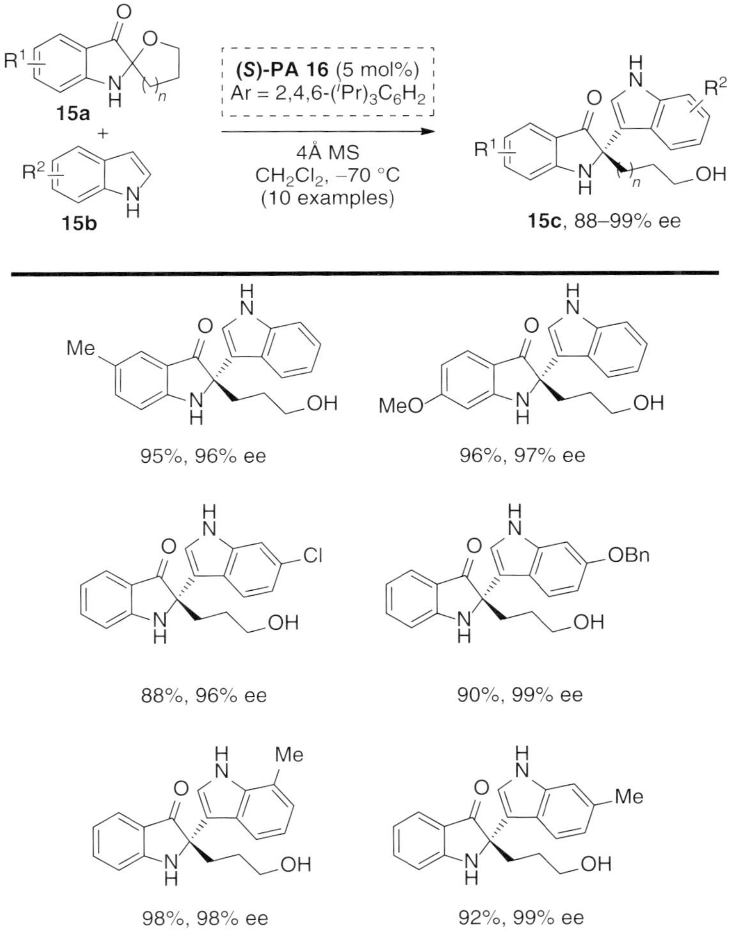

Scheme 3.28 Friedel–Crafts reaction of racemic spiroindolinones by You [36].

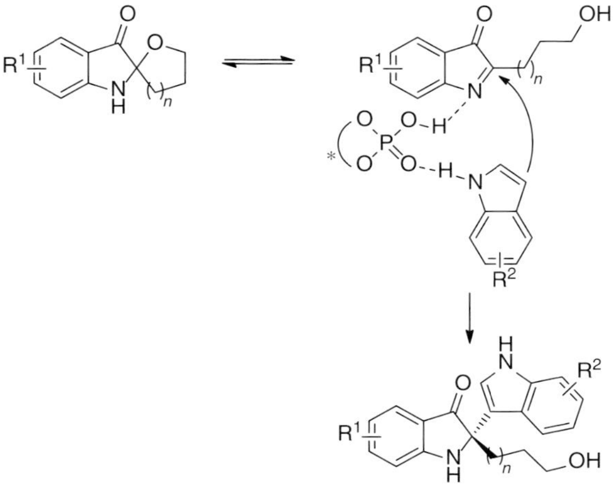

Scheme 3.29 Mechanism of Friedel–Crafts reaction [36].

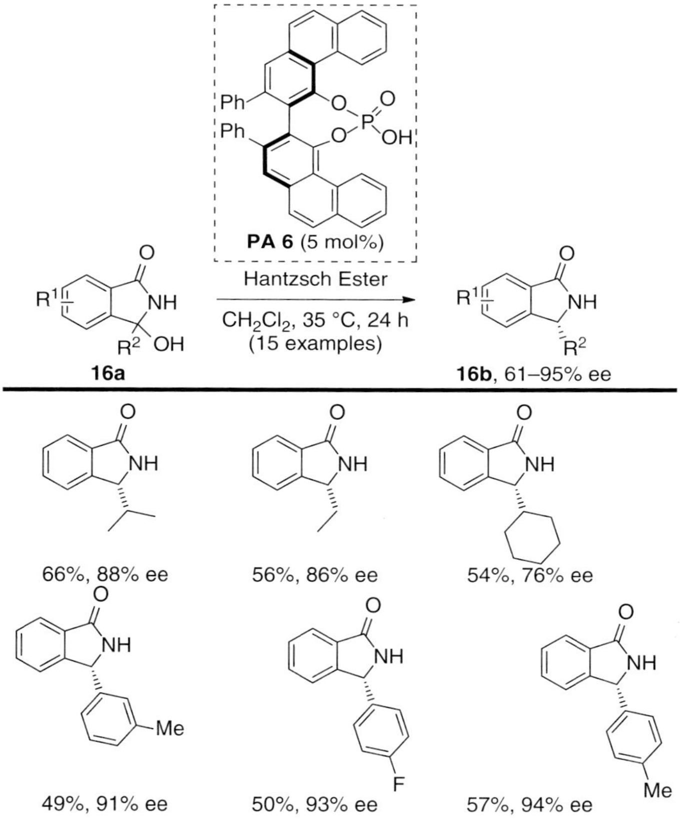

Scheme 3.30 Transfer hydrogenation of *N,O*-acetals by Zhou [37].

3.4
Miscellaneous Formation

This section is dedicated to a variety of substrates that can act as imine or iminium precursors but do not fall into the previously described categories. In each case, usually only one or two reports of each type exist, but this should not detract any attention away from these protocols as they serve as highly useful alternatives and may also provide unforeseen benefits. A rather interesting report was published by Akiyama in 2011, concerning an internal redox process that generated electrophilic iminium intermediates from benzyl protected anilines **17a** (Scheme 3.31) [40].

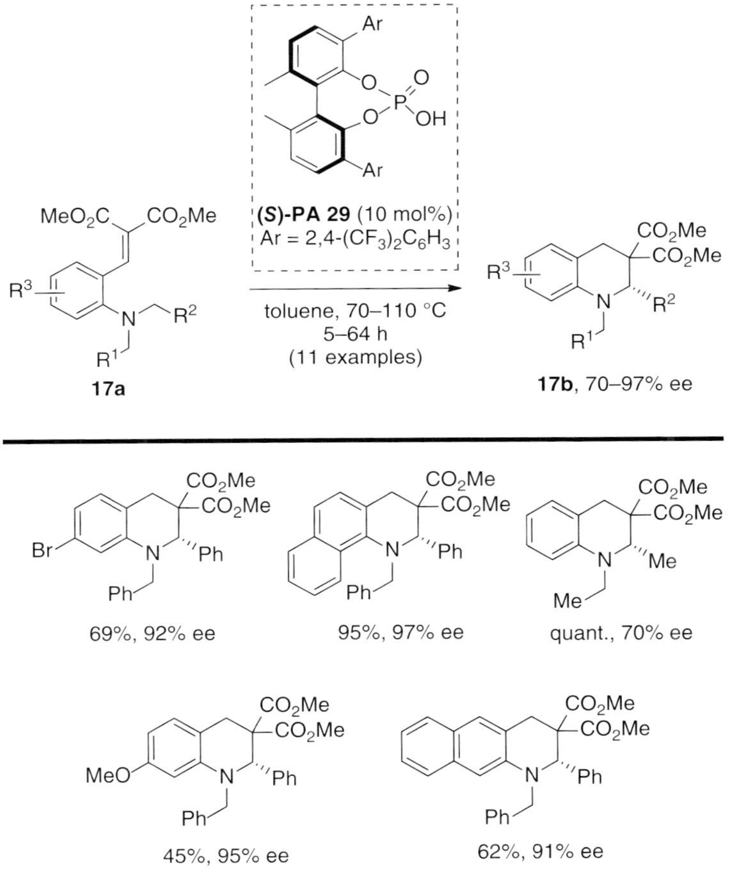

Scheme 3.31 Activation of enantiotopic C(sp^3)-hydrogen atoms by Akiyama [40].

It is proposed that treatment of **17a** with biphenyl-derived catalyst **(*S*)-PA 29** results in an internal hydride shift. The mechanism of this is given in Scheme 3.32.

After the hydride shift, the generated enolate can subsequently cyclize onto the iminium intermediate to deliver the tetrahydroquinoline products. The origin of selectivity is actually thought to be due to enantiotopic hydrogen differentiation by the catalyst.

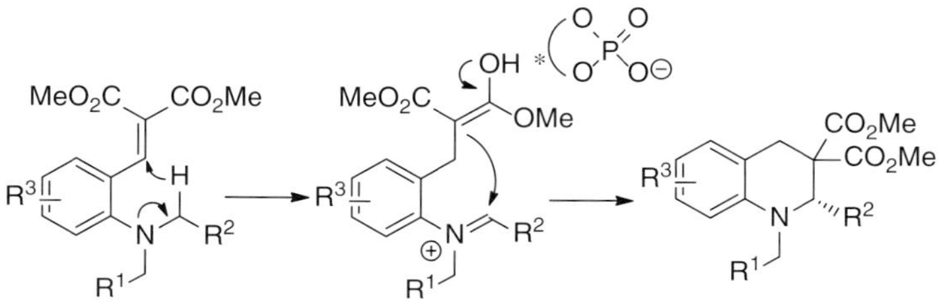

Scheme 3.32 Mechanism of hydride shift cyclization [40].

We have encountered N-acyl cyclic iminium ions previously within this book in a number of applications; however, they have generally been formed by the elimination of a small molecule (e.g., OH) at the γ-position of lactams. In 2011, Huang disclosed a rather unique entry to access such intermediates from α,β-unsaturated lactams **18a** (Scheme 3.33) [41].

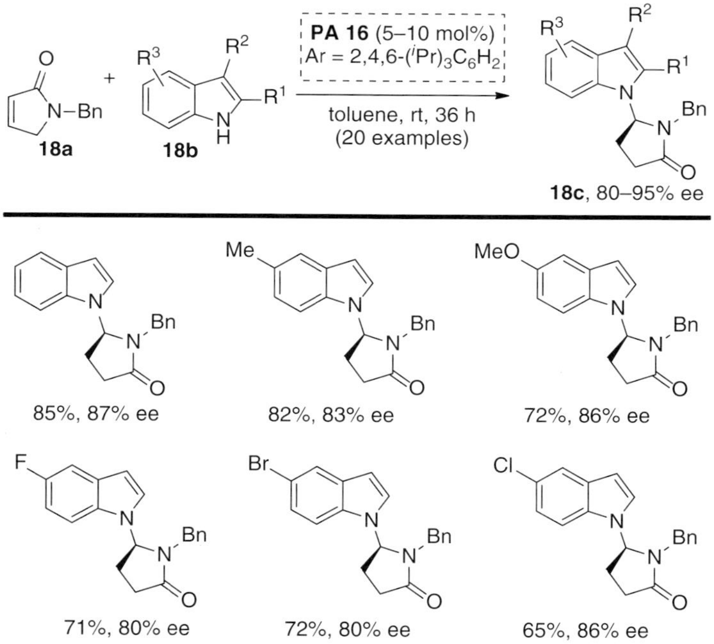

Scheme 3.33 Coupling of α,β-unsaturated lactams with indoles by Huang [41].

He showed that by treating **18a** with **PA 16**, a Friedel–Crafts reaction occurred with indoles **18b** to yield the adducts **18c** with good levels of selectivity. The mechanism for this reaction is proposed in Scheme 3.34. It is thought that isomerization of the starting material occurs to generate an *N*-acyliminium ion containing a bound chiral phosphate. The isomerization process was supported by deuterium-labeling studies. This substrate may exist as an isolated ion pairing, or there may also be hydrogen bonding between the enol tautomer. The authors propose that the coordination to the acidic proton of the indole renders the nitrogen sufficiently nucleophilic to attack directly onto the carbon to yield the desired products.

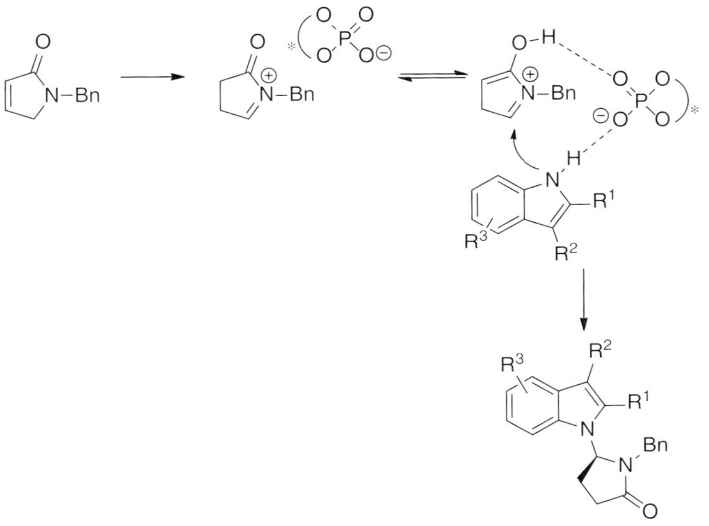

Scheme 3.34 Mechanism of coupling of α,β-unsaturated lactams with indoles [41].

The Toste group has been one of the pioneers in advancing the use of chiral Brønsted acid catalysis, and in 2013, they reported a novel combination of an oxidant (**19b**) with a specially designed phosphoric acid **PA 30** to perform cross-dehydrogenative couplings (Scheme 3.35) [42].

By treatment of amide-bound tetrahydroisoquinolines **19a** with **19b** in the presence of **PA 30** resulted in an oxidative cyclization of the amide onto the tetrahydroisoquinoline ring to yield the corresponding products **19c** with high enantioselectivities.

The mechanism of the process is depicted in Scheme 3.36. It begins by anion metathesis of the oxidant by the catalyst, which goes on to oxidize the starting material to an iminium ion. The iminium intermediate contains the bound phosphate catalyst, which is able to control the subsequent cyclization of the pendant amide group. It is proposed that the 3,3′-substituents on the catalyst also play the role of coordinating with the starting material to aid in the enantioselective step. A related procedure involving the pairing of a chiral phosphate with an oxidant has also been reported by Toste to access enantiopure indolines [43].

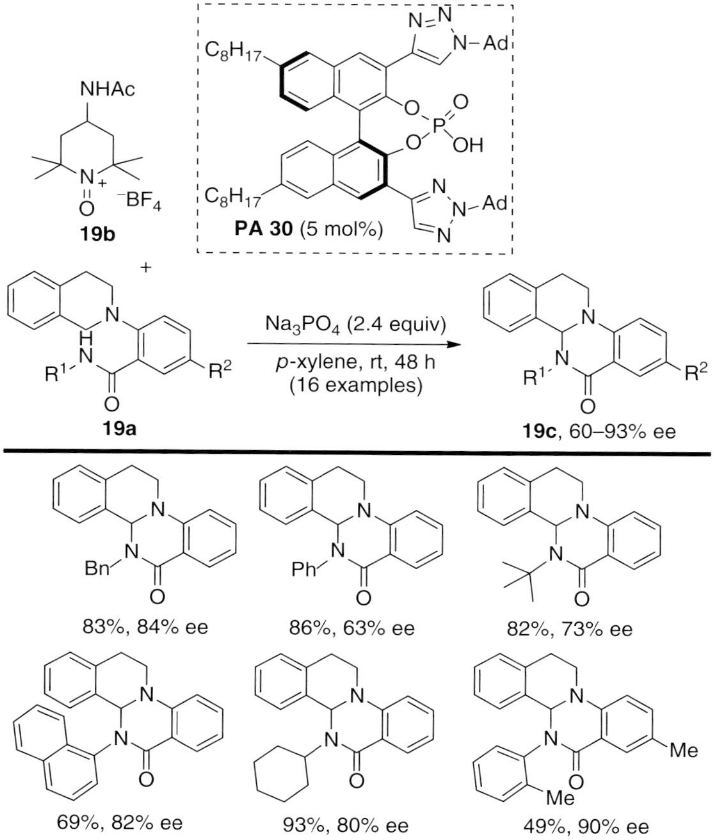

Scheme 3.35 Asymmetric cross-dehydrogenative coupling by Toste [42].

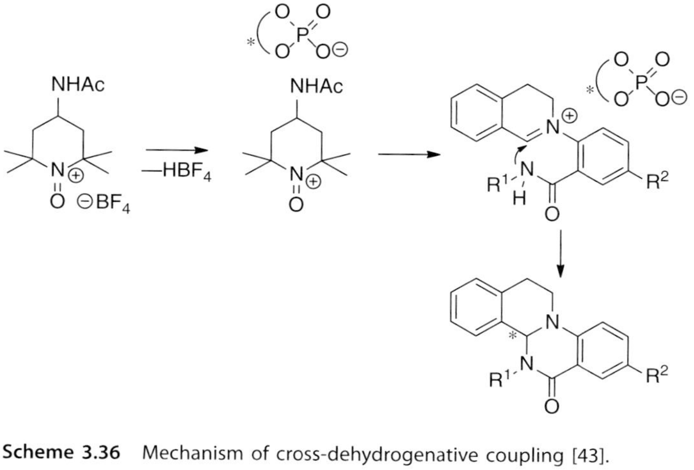

Scheme 3.36 Mechanism of cross-dehydrogenative coupling [43].

References

1. Baudequin, C., Zamfir, A., and Tsogoeva, S.B. (2008) *Chem. Commun.*, **2008**, 4637–4639.
2. Terada, M., Tanaka, H., and Sorimachi, K. (2008) *Synlett*, (11), 1661–1664.
3. Terada, M. and Sorimachi, K. (2007) *J. Am. Chem. Soc.*, **129**, 292–293.
4. Jia, Y.-X., Zhong, J., Zhu, S.-F., Zhang, C.-M., and Zhou, Q.-L. (2007) *Angew. Chem. Int. Ed.*, **46**, 5565–5567.
5. Romano, C., Jia, M., Monari, M., Manoni, E., and Bandini, M. (2014) *Angew. Chem. Int. Ed.*, **53**, 13854–13857.
6. Wang, S.-G. and You, S.-L. (2014) *Angew. Chem. Int. Ed.*, **53**, 2194–2197.
7. Li, G. and Antilla, J.C. (2009) *Org. Lett.*, **11**, 1075–1078.
8. Rueping, M. and Antonchick, A.P. (2008) *Angew. Chem. Int. Ed.*, **47**, 5836–5838.
9. Guo, Q.-X., Peng, Y.-G., Zhang, J.-W., Song, L., Feng, Z., and Gong, L.-Z. (2009) *Org. Lett.*, **11**, 4620–4623.
10. Guo, C., Song, J., Huang, J.-Z., Chen, P.-H., Luo, S.-W., and Gong, L.-Z. (2012) *Angew. Chem. Int. Ed.*, **51**, 1046–1050.
11. Song, L., Guo, Q.-X., Li, X.-C., Tian, J., and Peng, Y.-G. (2012) *Angew. Chem. Int. Ed.*, **51**, 1899–1902.
12. Liu, Y., Zhang, H.-H., Zhang, Y.-C., Jiang, Y., Shi, F., and Tu, S.-J. (2014) *Chem. Commun.*, **50**, 12054–12057.
13. Dai, W., Lu, H., Li, X., Shi, F., and Tu, S.-J. (2014) *Chem. Eur. J.*, **20**, 11382–11389.
14. Tan, W., Li, X., Gong, Y.-X., Ge, M.-D., and Shi, F. (2014) *Chem. Commun.*, **50**, 15901–15904.
15. Shi, F., Zhang, H.-H., Sun, X.-X., Liang, J., Fan, T., and Tu, S.-J. (2015) *Chem. Eur. J.*, **21**, 3465–3471.
16. Liang, T., Zhang, Z., and Antilla, J.C. (2010) *Angew. Chem. Int. Ed.*, **49**, 9734–9736.
17. Wu, X.-S. and Tian, S.-K. (2012) *Chem. Commun.*, **48**, 898–900.
18. Terada, M., Moriya, K., Kanomata, K., and Sorimachi, K. (2011) *Angew. Chem. Int. Ed.*, **50**, 12586–12590.
19. Sun, F.-L., Zheng, X.-J., Gu, Q., He, Q.-L., and You, S.-L. (2010) *Eur. J. Org. Chem.*, **2010**, 47–50.
20. Zhou, L.-J., Zhang, Y.-C., Zhao, J.-J., Shi, F., and Tu, S.-J. (2014) *J. Org. Chem.*, **79**, 10390–10398.
21. Zhuo, M.-H., Jiang, Y.-J., Fan, Y.-S., Gao, Y., Liu, S., and Zhang, S. (2014) *Org. Lett.*, **16**, 1096–1099.
22. Qi, S., Liu, C.-Y., Ding, J.-Y., and Han, F.-S. (2014) *Chem. Commun.*, **50**, 8605–8608.
23. Sun, F.-L., Zeng, M., Gu, Q., and You, S.-L. (2009) *Chem. Eur. J.*, **15**, 8709–8712.
24. Wang, S.-G., Han, L., Zeng, M., Sun, F.-L., Zhang, W., and You, S.-L. (2012) *Org. Biomol. Chem.*, **10**, 3202–3209.
25. Terada, M., Machioka, K., and Sorimachi, K. (2009) *Angew. Chem. Int. Ed.*, **48**, 2553–2556.
26. Terada, M. and Toda, Y. (2009) *J. Am. Chem. Soc.*, **131**, 6354–6355.
27. Terada, M., Komuro, T., Toda, Y., and Korenaga, T. (2014) *J. Am. Chem. Soc.*, **136**, 7044–7057.
28. Zhang, H.-H., Wang, Y.-M., Xie, Y.-W., Zhu, Z.-Q., Shi, F., and Tu, S.-J. (2014) *J. Org. Chem.*, **79**, 7141–7151.
29. Kano, T., Yurino, T., and Maruoka, K. (2013) *Angew. Chem. Int. Ed.*, **52**, 11509–11512.
30. Rueping, M. and Nachtsheim, B.J. (2010) *Synlett*, (1), 119–122.
31. Aranzamendi, E., Sotomayor, N., and Lete, E. (2012) *J. Org. Chem.*, **77**, 2986–2991.
32. Gómez-SanJuan, A., Sotomayor, N., and Lete, E. (2012) *Tetrahedron Lett.*, **53**, 2157–2159.
33. Yu, X., Lu, A., Wang, Y., Wu, G., Song, H., Zhou, Z., and Tang, C. (2011) *Eur. J. Org. Chem.*, **2011**, 892–897.
34. Yu, X., Wang, Y., Wu, G., Song, H., Zhou, Z., and Tang, C. (2011) *Eur. J. Org. Chem.*, **2011**, 3060–3066.
35. Courant, T., Kumarn, S., He, L., Retailleau, P., and Masson, G. (2013) *Adv. Synth. Catal.*, **355**, 836–840.
36. Yin, Q. and You, S.-L. (2011) *Chem. Sci.*, **2**, 1344–1348.
37. Chen, M.-W., Chen, Q.-A., Duan, Y., Ye, Z.-S., and Zhou, Y.-G. (2012) *Chem. Commun.*, **48**, 1698–1700.

38. Yin, Q., Wang, S.-G., and You, S.-L. (2013) *Org. Lett.*, **15**, 2688–2691.
39. Zhou, J.-Q., Sheng, W.-J., Jia, J.-H., Ye, Q., Gao, J.-R., and Jia, Y.-X. (2013) *Tetrahedron Lett.*, **54**, 3082–3084.
40. Mori, K., Ehara, K., Kurihara, K., and Akiyama, T. (2011) *J. Am. Chem. Soc.*, **133**, 6166–6169.
41. Xie, Y., Zhao, Y., Qian, B., Yang, L., Xia, C., and Huang, H. (2011) *Angew. Chem. Int. Ed.*, **50**, 5682–5686.
42. Neel, A.J., Hehn, J.P., Tripet, P.F., and Toste, F.D. (2013) *J. Am. Chem. Soc.*, **135**, 14044–14047.
43. Lackner, A.D., Samant, A.V., and Toste, F.D. (2013) *J. Am. Chem. Soc.*, **135**, 14090–14093.

4
Reactions of Carbonyls

Carbonyl substrates are considered as one of the toughest challenges for Brønsted acid activations due to their significantly lower basicity compared to imine substrates. Nevertheless, research groups have overcome this challenge and have shown that highly useful methodologies can be developed. Despite this, the balance of research papers is still swayed in the favor of imine substrates, and, therefore, the activation of carbonyl substrates still represents an area for potential new advances.

4.1
Nucleophilic Addition Reactions

4.1.1
C-Nucleophiles

The addition of C-nucleophiles represents an important category, since the creation of C–C bonds in an asymmetric fashion is a highly useful process. The following examples represent the most common types of C-nucleophiles such as allyl-reagents or enols, but more obscure substrates such as isocyanoacetamides have been developed in an isolated example [1]. The asymmetric allylation of aldehydes is an important reaction to the synthetic community, which has attracted high levels of attention from research groups employing a variety of chiral catalysts. In 2010, Antilla disclosed results that showed that a phosphoric acid catalyst (**PA 16**) was able to catalyze the coupling of aldehydes **1a** and allyl boronates **1b** to yield the adducts **1c** with high levels of enantioselectivity (Scheme 4.1) [2].

The reaction generally worked well for aromatic aldehydes and gave slightly lower levels of selectivity for aliphatic substrates. The reaction could also be extended to crotylations with equally impressive results. In 2012, Hu employed a SPINOL-derived catalyst (**SPA 1**), which was shown to provide similar products **1f** with high enantioselectivities (Scheme 4.2) [3].

The mechanism of this reaction has been hotly investigated by the groups of Houk [4] and Goodman [5], who respectively provide evidence for the major

Asymmetric Brønsted Acid Catalysis, First Edition. Magnus Rueping, Dixit Parmar and Erli Sugiono.

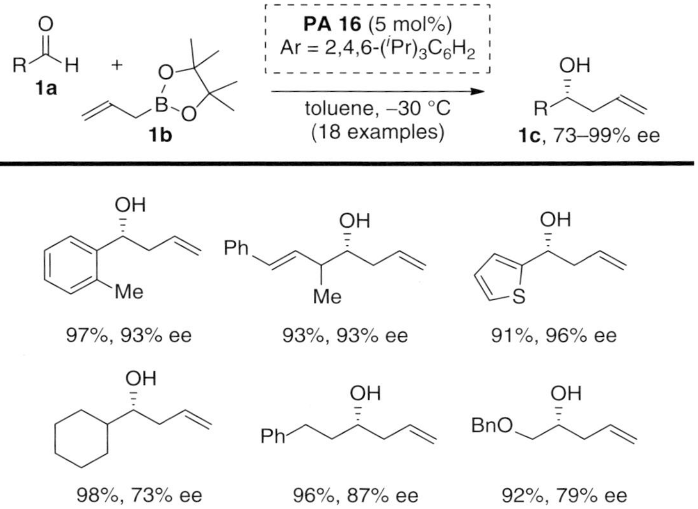

Scheme 4.1 Allylboration of aldehydes by Antilla [2].

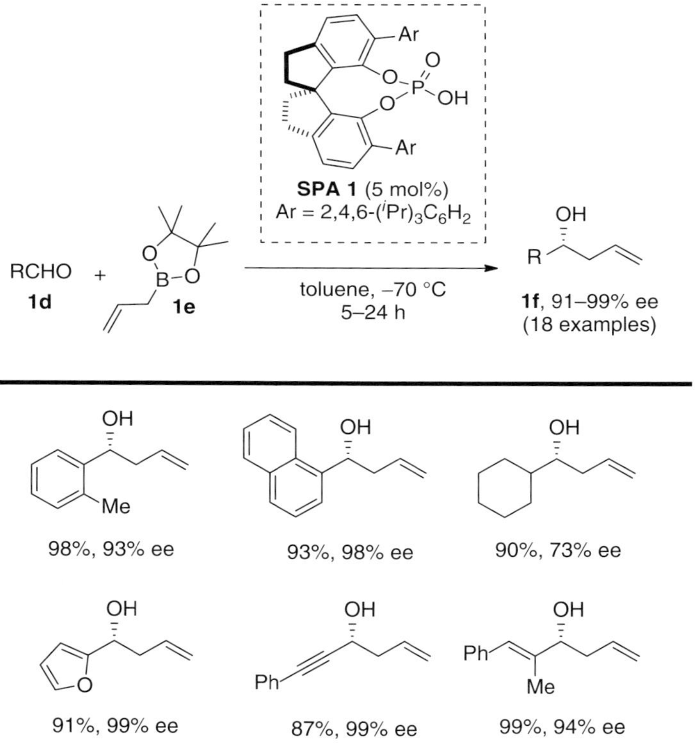

Scheme 4.2 Allylboration of aldehydes by Hu [3].

and minor transition states (Scheme 4.3). Both groups carried out extensive computational calculations, which fit the experimental results obtained from Antilla very well. The major point of difference is whether the catalyst interacts with the pseudoaxial oxygen atom or the pseudoequatorial oxygen atom. More recently, other research groups have published reports on allyl boronates with aldehydes by the use of chiral phosphoric acid catalysts [6, 7]. Antilla has also extended his work to the propargylation of aldehydes. The treatment of aldehydes **2a** with allenylboronic ester **2b** in the presence of 20 mol% **PA 16** resulted in a stereoselective addition to give the products **2c** with good to high selectivities (Scheme 4.4) [8].

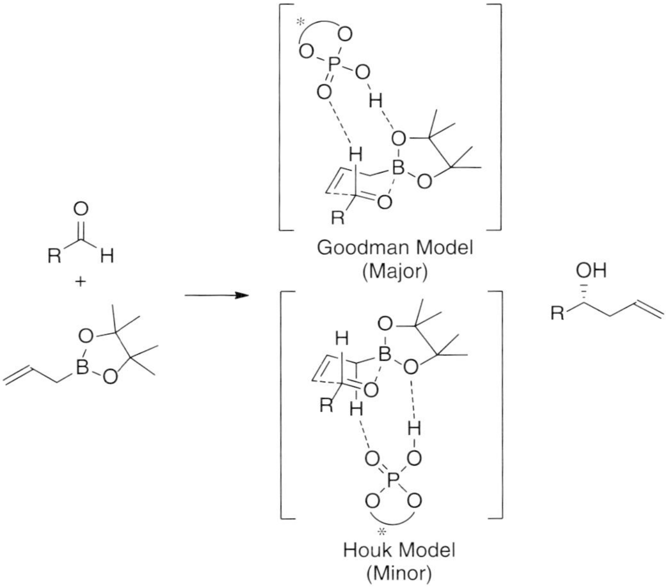

Scheme 4.3 Mechanism of allylboration of aldehydes by Goodman and Houk [4, 5].

The reaction delivers similar results to the allylation, whereby aromatic substrates provide higher levels of enantioselectivity than aliphatic substrates. This reaction has also been studied by Roush in performing kinetic resolutions of allenylboronates (±)-**2e** by coupling with aldehydes **2d** in the presence of 10 mol% **PA 16** (Scheme 4.5) [9, 10]. The mechanism of this process has been studied by the Goodman group, which presents a reaction model that correctly predicts the stereochemical outcome of the process [11]. Reddy has also studied propargylations and allenylations using chiral phosphoric acid catalysts [12, 13].

Allylation-type products are useful building blocks since they allow synthetic chemists to perform a whole host of further transformations. Research groups have also studied routes to allylation products via different reactive intermediates. In 2008, Rueping developed the asymmetric ene reaction between α-methyl

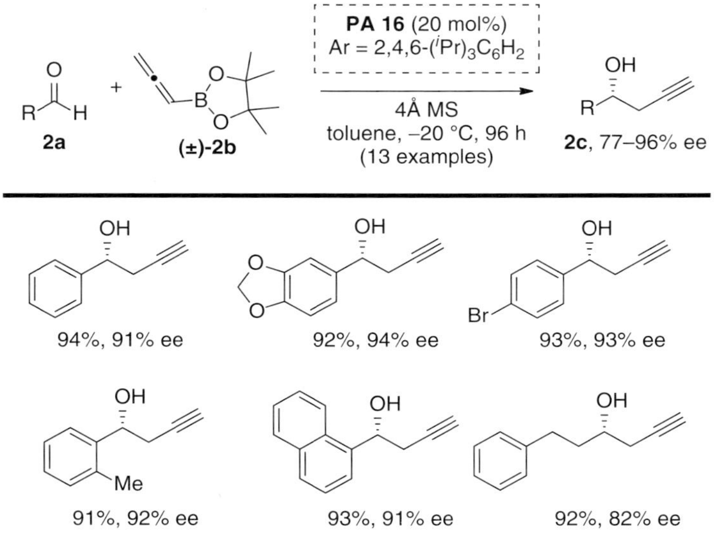

Scheme 4.4 Propargylation of aldehydes by Antilla [8].

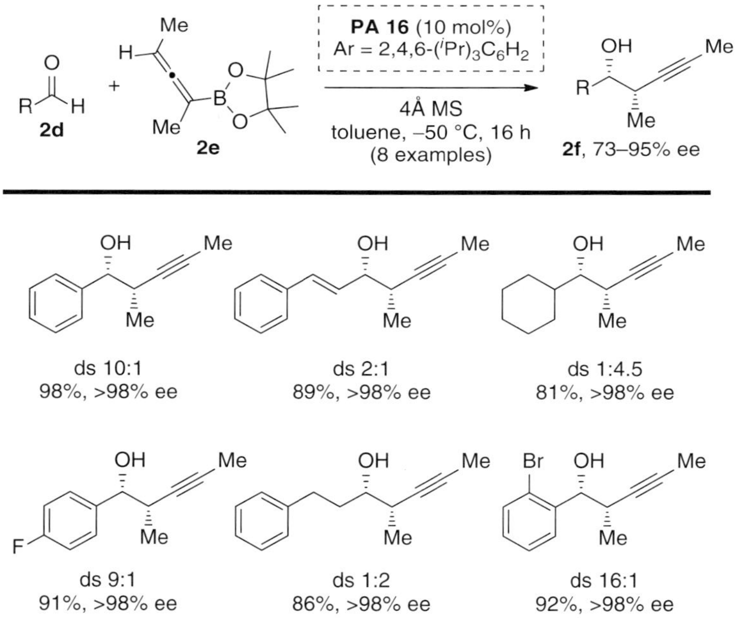

Scheme 4.5 Propargylation of aldehydes by Roush [9].

styrenes **3a** and trifluoropyruvate **3b** to yield the allylation-type products **3c** (Scheme 4.6) [14].

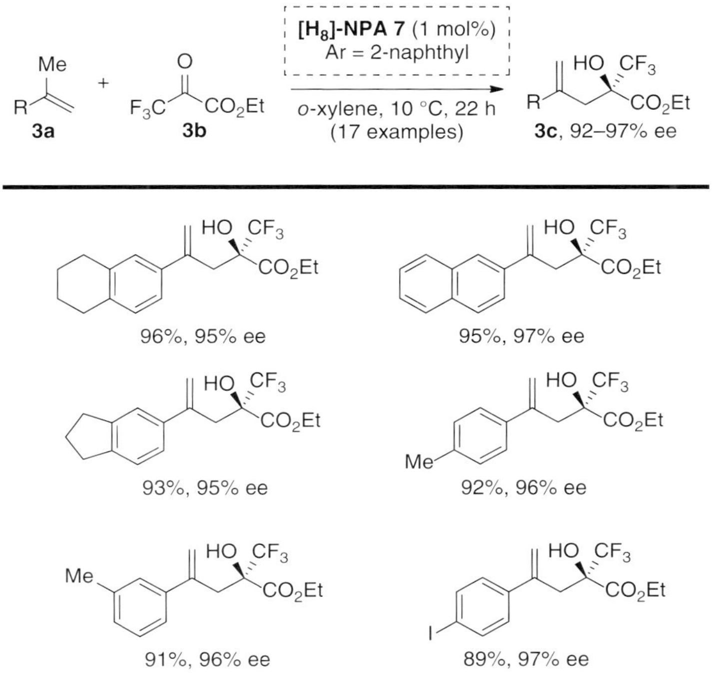

Scheme 4.6 Carbonyl-ene reaction by Rueping [14].

In this instance, the more acidic *N*-triflylphosphoramide catalyst **[H$_8$]-NPA 7** was required for activation; however, just 1 mol% provided excellent levels of enantioselectivity. The aza-ene reaction was also published in the same year by Terada [15]. In 2012, List reported the allylation of aldehydes **3d** via the Hosomi–Sakurai reaction with various allyl silanes **3e** using 10 mol% of disulfonimide catalyst (**BA 12**) (Scheme 4.7) [16]. The products **3f** were generally obtained with high selectivities; however, the use of unsubstituted allyl silanes (**3e**, $R^2 = H$) gave poor results.

The mechanism of the carbonyl-ene reaction is depicted in Scheme 4.8. Activation of the ketone occurs by the more acidic *N*-triflylphosphoramide catalyst, which allows for the alkene to undergo a most likely concerted attack and loss of a proton to yield the products and regenerate the catalyst. The mechanism of the Hosomi–Sakurai reaction is proposed to proceed in a similar fashion as illustrated in Scheme 2.11.

In 2009, Akiyama developed a Brønsted acid approach to chiral cyclohexanones, which would rival the use of more conventional covalent chiral catalysts such as proline. The reaction in question was a desymmetrization of meso-1,3-diones **4a**

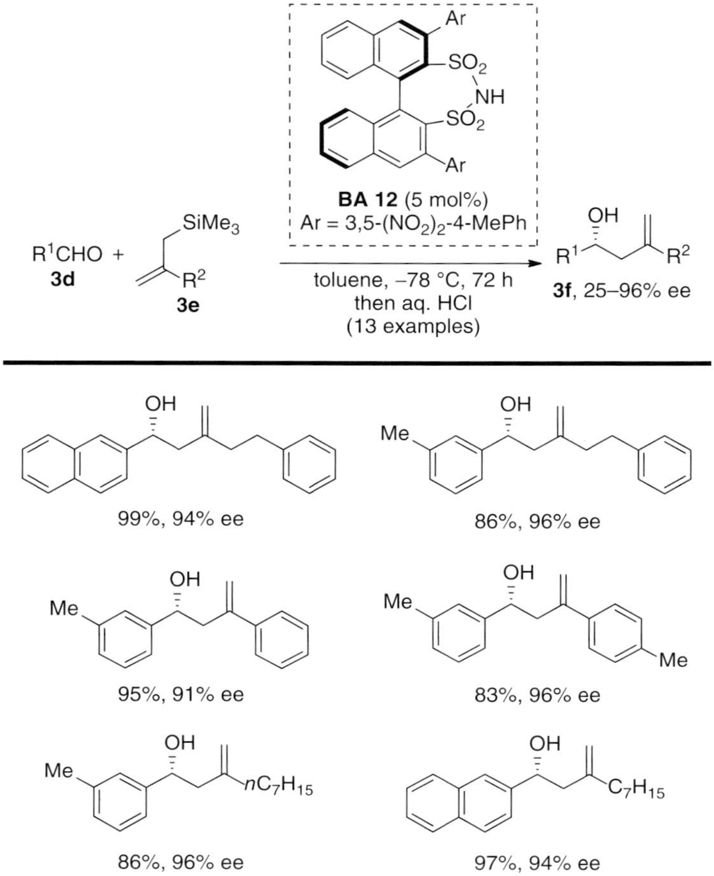

Scheme 4.7 Hosomi–Sakurai reaction by List [16].

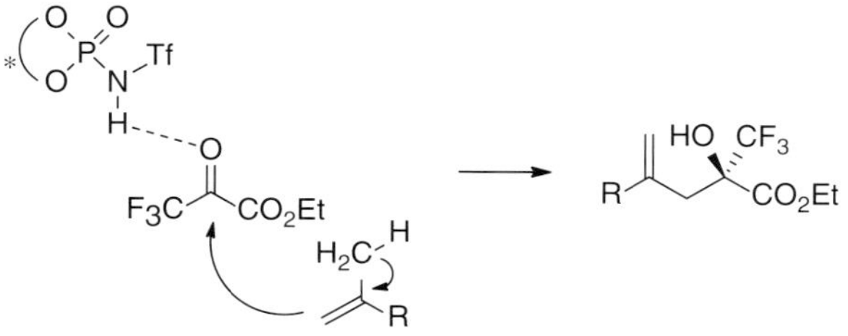

Scheme 4.8 Mechanism of carbonyl-ene reaction [14].

in a Robinson-like annulation reaction to access bicyclic products **4b** with high enantioselectivities (Scheme 4.9) [17–19].

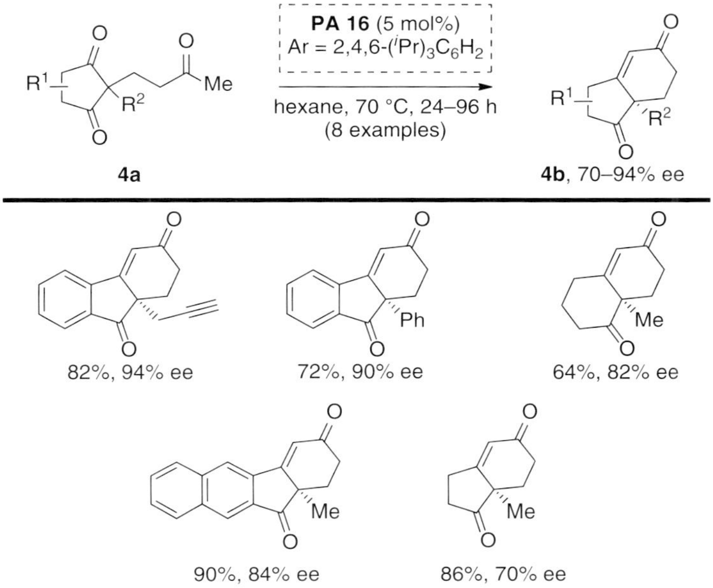

Scheme 4.9 Desymmetrization using a Robinson annulation by Akiyama [17].

In general, the reaction proceeded well for a series of aromatic and aliphatic substituted substrates. The mechanism of the reaction is illustrated in Scheme 4.10. It is proposed that the catalyst activates both the carbonyl and enol functionalities, which allows for selective attack to occur at one of the carbonyl groups. Dehydration of the intermediate formed yields the desired products. ONIOM calculations were also performed, which support the activation mode and of attack on the *si* face.

Finally, in this subsection, we cover Friedel–Crafts reactions, which for other substrates such as imines have a separate section dedicated to them, but for carbonyl substrates, very few reports exist for their successful use. Ma's group has been the most prominent in this area, and in 2009, they disclosed their results on the Friedel–Crafts reactions of N–H-free indoles **5a** with trifluoroketones **5b** to give the adducts **5c** generally with high enantioselectivities (Scheme 4.11, top) [20, 21]. Akiyama has also disclosed one report regarding this reaction [22].

In 2010, Ma extended his protocol to include the use of 4,7-dihydroindoles **5d** with trifluoroketones **5e** and catalyst **PA 31** to give the adducts **5f** with good selectivities (Scheme 4.11, bottom) [23]. Although the CF_3 is an attractive feature of these substrates, the reaction has not been shown to be general for all ketone substrates.

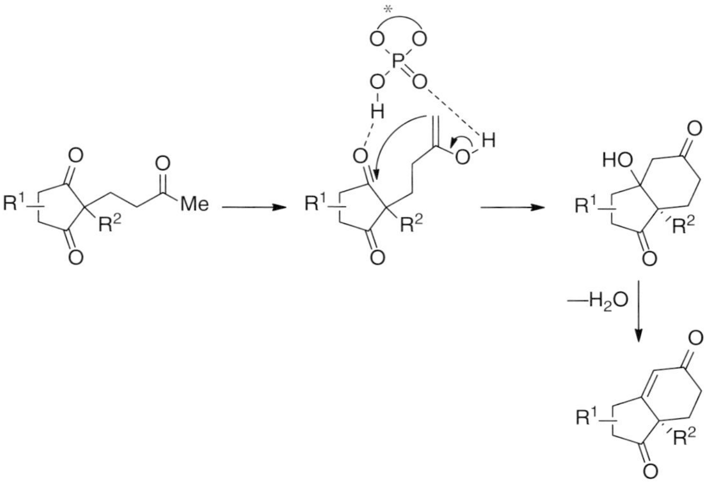

Scheme 4.10 Mechanism of desymmetrization [17].

The mechanisms for both Friedel–Crafts reactions are shown in Scheme 4.12. In both cases, the catalyst is thought to activate both components during the stereoselective addition step. In the first case with indoles, attack occurs at the C3-position, while with 4,7-dihydroindoles, attack occurs at the C2-position. With the 4,7-dihydroindoles being easily oxidized to the indoles, the two protocols are highly complementary as they allow access to either 2- or 3-substituted indoles.

4.1.2
N-, O-, and P-Nucleophiles

Heteroatom-nucleophiles represent a significant quantity of the research into reactions with carbonyl substrates as the products function as highly versatile intermediates or possess biological activities. In comparing nitrogen and oxygen nucleophiles, oxygen nearly accommodates the entire literature as basic nitrogen nucleophiles can possess challenges of compatibility with acidic catalysts. In an isolated report, Zhang has found that the stereoselective addition of nitrogen to a carbonyl could be carried out by the use of alkyl azides **6a** in combination with a super stoichiometric amount of **NPA 4** (Scheme 4.13) [24].

Formally, the transformation can be thought of as an intramolecular Schmidt reaction performed as a desymmetrization. Although the reaction only delivers modest enantioselectivities of the products **6b**, the reaction represents a breakthrough for the field.

The mechanism for the process is portrayed in Scheme 4.14. It is proposed that the catalyst activates selectively one carbonyl group while also providing additional coordination to the positively charged azide group. This promotes the subsequent attack, and then this is followed by a rearrangement process and the loss of N_2 to yield the lactam products. A desymmetrizing lactamization has also recently

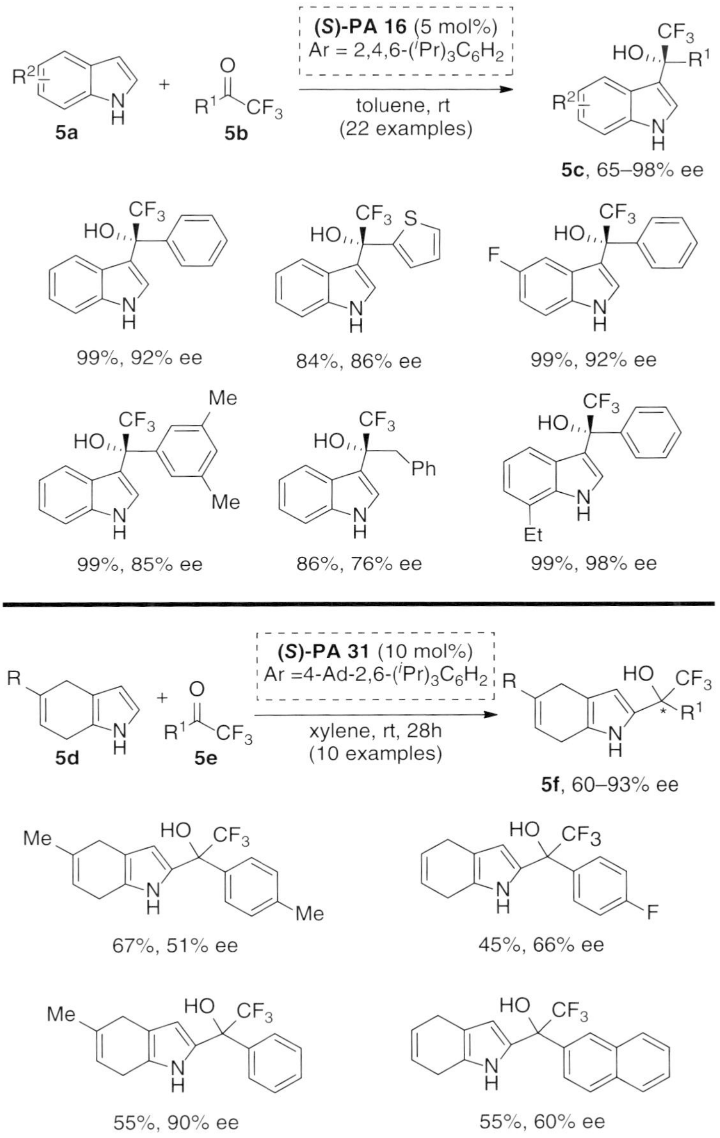

Scheme 4.11 Friedel–Crafts reactions of ketones by Ma [20, 21, 23].

been employed in the total synthesis of (−)-leuconoxine by Higuchi and Kawasaki [25]. Desymmetrizations of carbonyls with oxygen nucleophiles have provided more pleasing results. In 2010, List developed a novel bifunctional catalyst ***(S)*-BA 13,** which was shown to facilitate the desymmetrization of anhydrides **7a** with MeOH as a nucleophile (Scheme 4.15) [26]. The catalyst was specially designed to

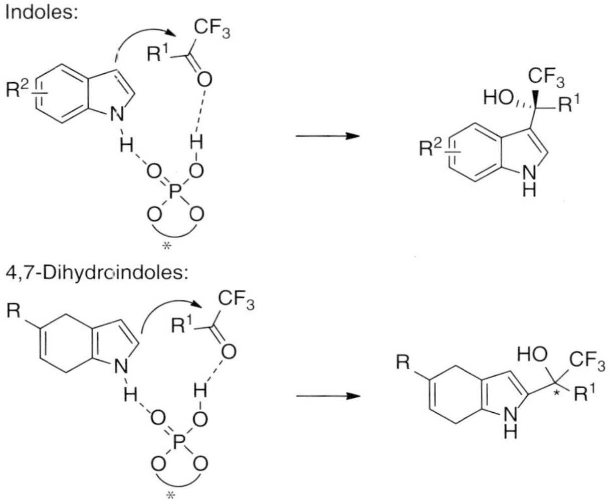

Scheme 4.12 Mechanism of Friedel–Crafts reactions [20, 21, 23].

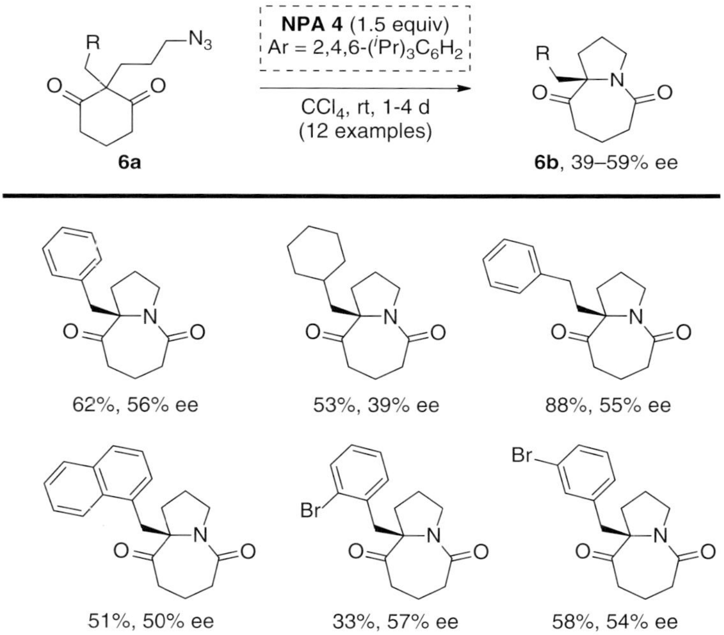

Scheme 4.13 Intramolecular Schmidt reaction by Zhang [24].

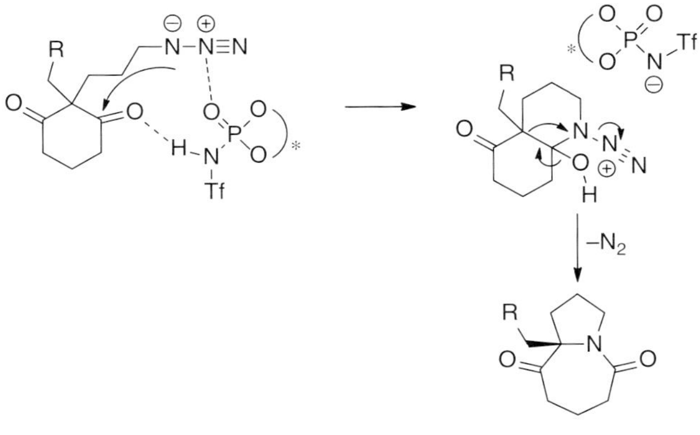

Scheme 4.14 Mechanism of Schmidt reaction [24].

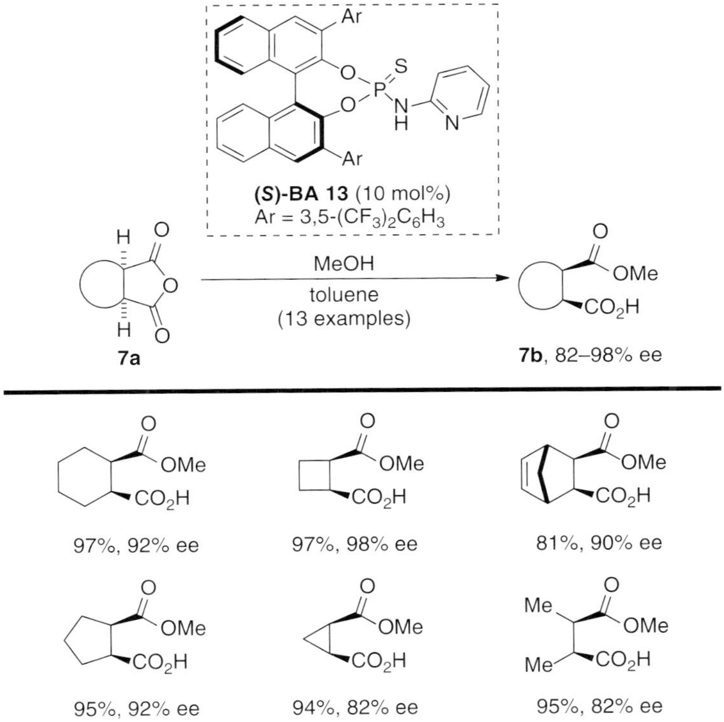

Scheme 4.15 Desymmetrization by List [26].

contain a Brønsted acidic site for activating the anhydride and a Brønsted basic site for activating the MeOH nucleophile.

The reaction shows a good scope and provides the products **7b** with generally high enantioselectivities. The protocol was also demonstrated to function well with other alcohols, and in addition, a short formal synthesis of a natural pheromone was reported. Recently, Zhu has developed a desymmetrization of bislactones **7c** with alcohol nucleophiles to access enantioenriched acids **7d** with high levels of enantioselectivity (Scheme 4.16) [27]. In his case, an imidodiphosphoric acid **(*S*)-NPA 8** was found to be the most optimal catalyst.

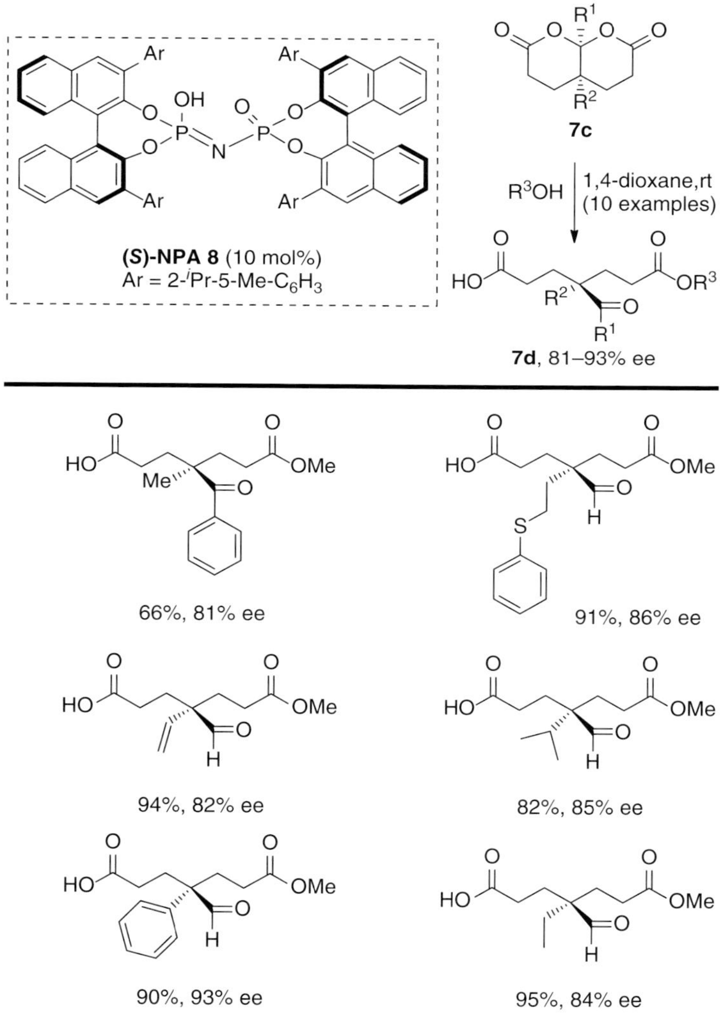

Scheme 4.16 Desymmetrization by Zhu [27].

The mechanism for the desymmetrization of bislactones is illustrated in Scheme 4.17. It can be envisioned that the catalyst activates both the alcohol nucleophile and selectively one carbonyl of the bislactone. Addition of the alcohol incurs a double ring-opening process to yield the desired products. Anhydrides have also been shown to be suitable for activation for use in the kinetic resolution of secondary alcohols. Yamada and Takasu showed in 2013 that isobutyric anhydride **8a** could be employed as an acylating agent for racemic alcohols **(±)-8b** to selectively form **8c** and leave behind enantioenriched **(−)-8b** in the presence of **PA 32** (Scheme 4.18) [28].

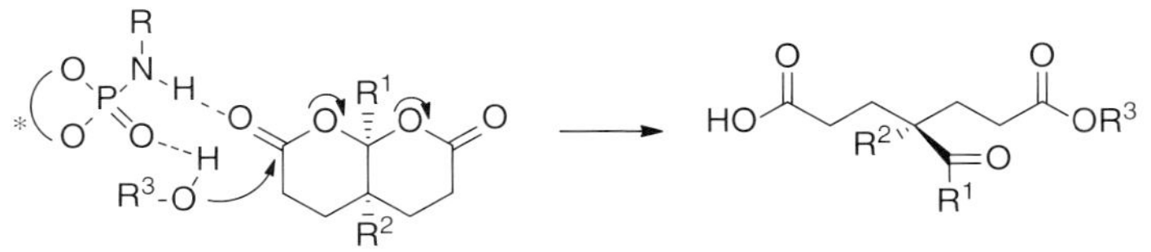

Scheme 4.17 Mechanism of desymmetrization [27].

This mechanism can be observed in Scheme 4.19. Detailed mechanistic experiments revealed that acylation of the catalyst is not a pathway in this transformation but instead a strong hydrogen-bond activation of the anhydride and coordination to the incoming alcohols is responsible for the stereoselectivity observed. The preference for one enantiomer of the alcohol over the other enantiomer results in the resolution process observed. The concept of kinetic resolution by activation of carbonyl substrates has also been explored by several other groups [29–32].

The Baeyer–Villiger reaction is one of the most essential tools in organic synthesis and has found utility in numerous applications over time. The asymmetric development of the reaction, however, has been challenging and progress has been rather moderate. In 2008, Ding reported a remarkable breakthrough for Brønsted acid catalysis in the successful activation of aqueous H_2O_2 to perform asymmetric Baeyer–Villiger oxidations. Ding realized that the treatment of cyclobutanones **9a** with aqueous H_2O_2 in the presence of 10 mol% **[H$_8$]-PA 33** delivered the corresponding lactone products **9b** with modest levels of enantioselectivity (Scheme 4.20) [33–35].

Generally, the reaction is extremely efficient and provides quantitative yields of the desired products. Interestingly, during the catalyst optimizations, octahydro-BINOL catalysts performed superior to BINOL-derived catalysts and **[H$_8$]-PA 33** was the only catalyst that provided acceptable results. The mechanism of the reaction is shown in Scheme 4.21. In the first step, activation of the carbonyl and the hydrogen peroxide occurs by the catalyst. The authors propose the mechanism involves preferential formation of a Criegee intermediate. Upon formation, the catalyst also promotes the stereoselective alkyl shift leading to the observed products.

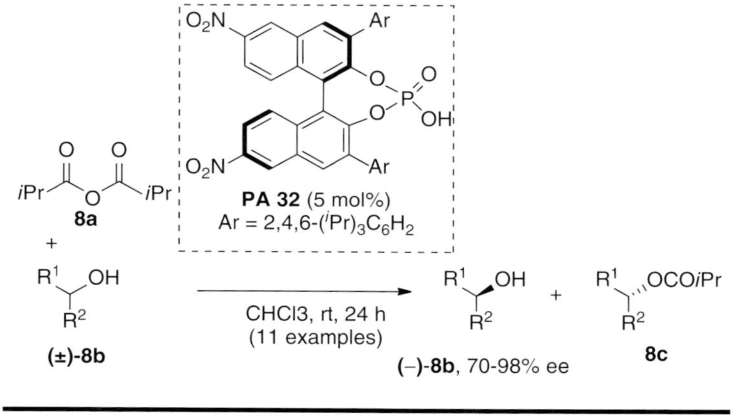

Scheme 4.18 Kinetic resolution of secondary alcohols by Yamada and Takasu [28].

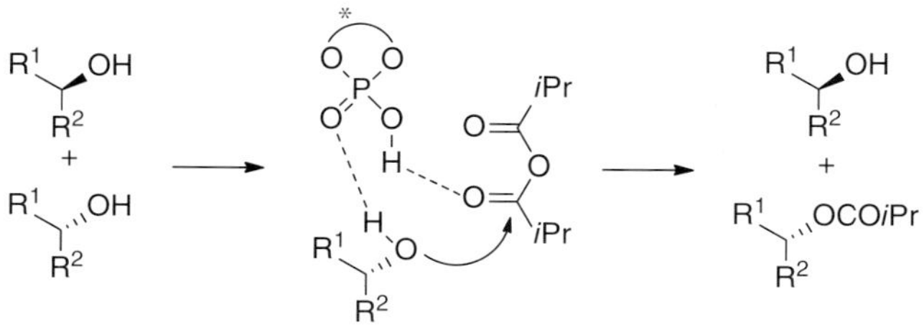

Scheme 4.19 Mechanism of kinetic resolution [28].

Acetal formation is a routine transformation known by synthetic chemists, usually as a means of protection for carbonyl species. Typically, the procedure is catalyzed by a Brønsted acid and therefore would serve as a suitable template for asymmetric formation. In 2013, List realized this opportunity and developed the asymmetric condensation of diols **10a** with aldehydes **10b** to yield enantioenriched acetals **10c** (Scheme 4.22) [36].

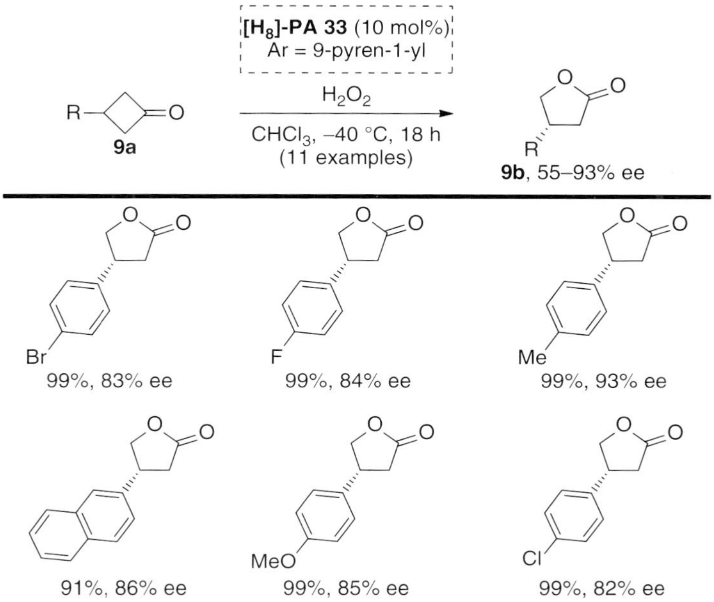

Scheme 4.20 Baeyer–Villiger reaction by Ding [33].

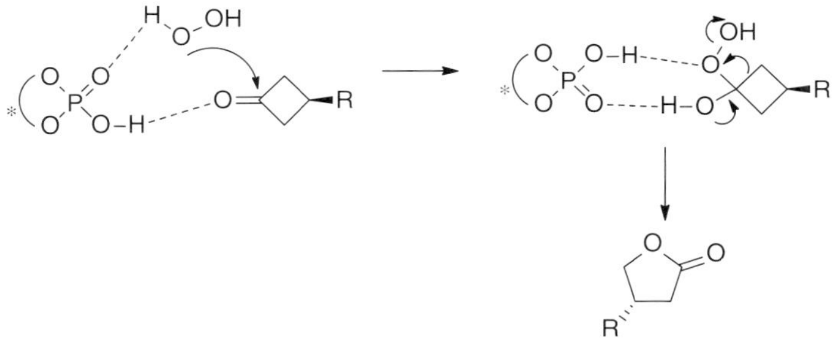

Scheme 4.21 Mechanism of Baeyer–Villiger reaction [33].

In line with the group's previously developed work, a C_2-symmetric imidodiphosphate **(*S*)-NPA 8** was employed to provide reasonable results; however, the substitution pattern on the aromatic group was found to be optimal only when unsymmetrical. The mechanism of the process is depicted in Scheme 4.23. It is proposed that elimination of H_2O from the starting material occurs, which leads to an oxocarbenium ion and a chiral imidodiphosphate counterion. This controls the stereoselective ring-closing acetalization process to give the desired products.

The addition of P-nucleophiles to imine substrates catalyzed by Brønsted acid catalysts is a well-studied area; however, the corresponding extension to carbonyl

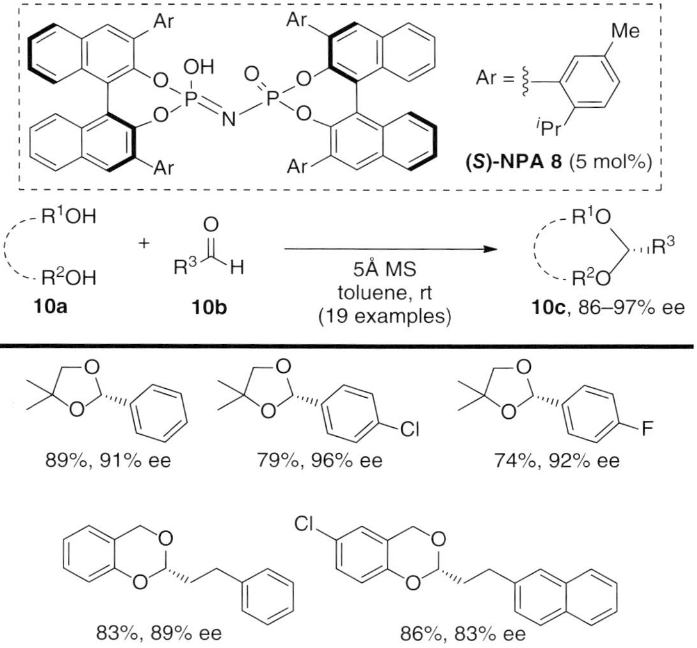

Scheme 4.22 Acetalization of aldehydes by List [36].

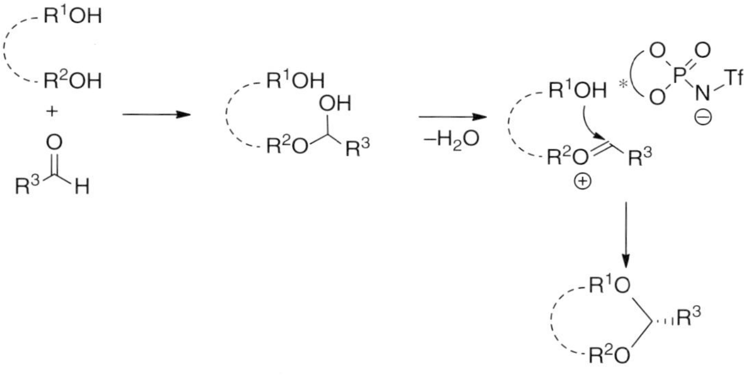

Scheme 4.23 Mechanism of acetalization [36].

substrates has remained unknown up until recently. List recognized that silyl-protected phosphites could be likened to silyl enol ethers and would undergo a nucleophilic addition process under appropriate conditions. He went on to show that phosphites **11a** could be coupled with aldehydes **11b** by using disulfonamide catalyst **BA 9** (Scheme 4.24) [37].

The reaction is sensitive to the phosphite substituents but could tolerate a wide variety of aromatic aldehydes. An alkyl aldehyde in contrast provided a modest

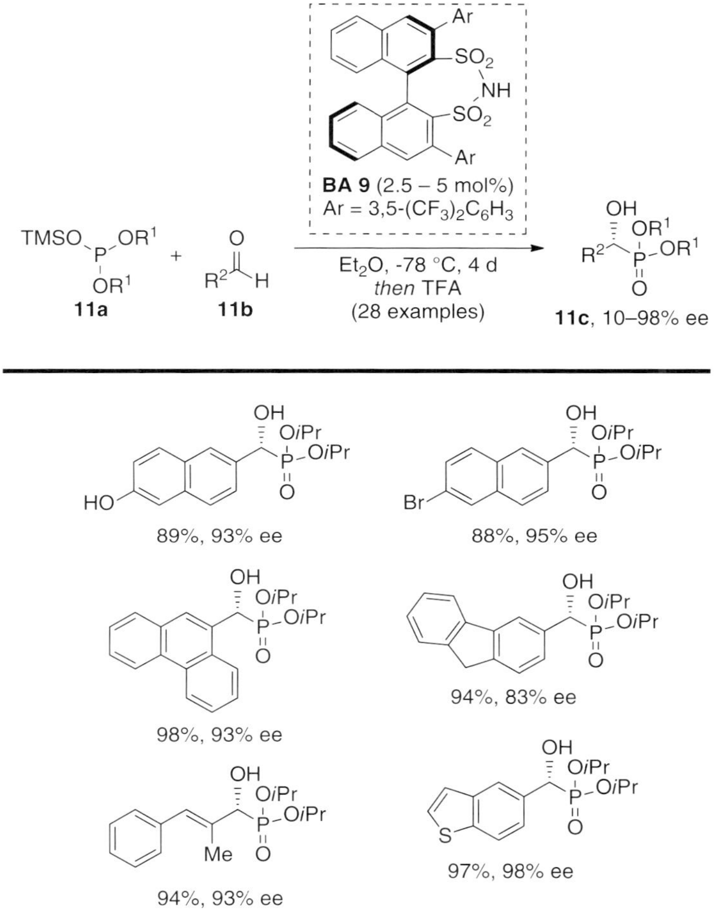

Scheme 4.24 Abramov reaction by List [37].

yield and almost no enantioselectivity. The products **11c** were obtained after a simple acidic hydrolysis following the initial addition process.

4.2
Aldol Reactions

The aldol reaction is perhaps one of the most famous reaction types in synthetic organic chemistry and has received attention from all corners of the community with regard to racemic and asymmetric developments. Chiral Brønsted acids have also played a part in this development and a handful of powerful reports have been disclosed. One of the first was by the List group, who reported a Mukaiyama aldol reaction between silyl ketene acetals **12a** and aldehydes **12b** in the presence of disulfonimide catalyst **BA 9** (Scheme 4.25) [38, 39].

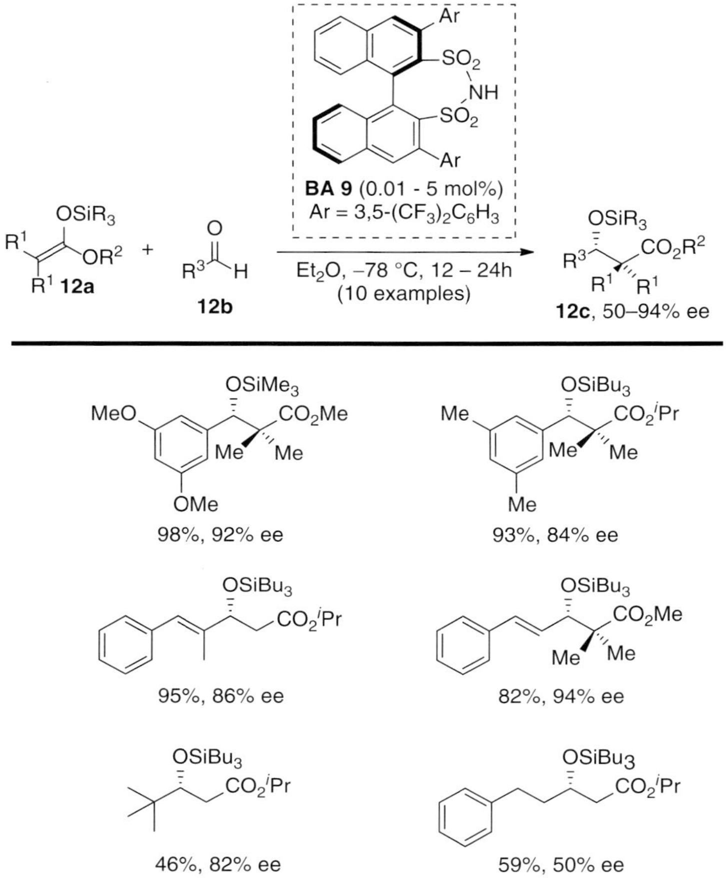

Scheme 4.25 Mukaiyama aldol reaction by List [38].

The reaction performs best with aromatic aldehydes; however, aliphatic aldehydes can be tolerated with modest results. The aldol products **12c** are obtained with good enantioselectivities. Yamamoto has also found similar results when using an *N*-triflylthiophosphoramide catalyst **(*S*)-BA 14,** which delivered the aldol products **12f** with generally high selectivities (Scheme 4.26) [40]. Addition of a base at room temperature failed to perturb the reaction, leading to the conclusion that the catalyst may perform in its silylated form.

The mechanism of the reaction is illustrated in Scheme 4.27. The authors propose that the catalyst actually provides a chiral counterion for a silylated-oxonium intermediate. So, the first stage of the process involves transfer of a

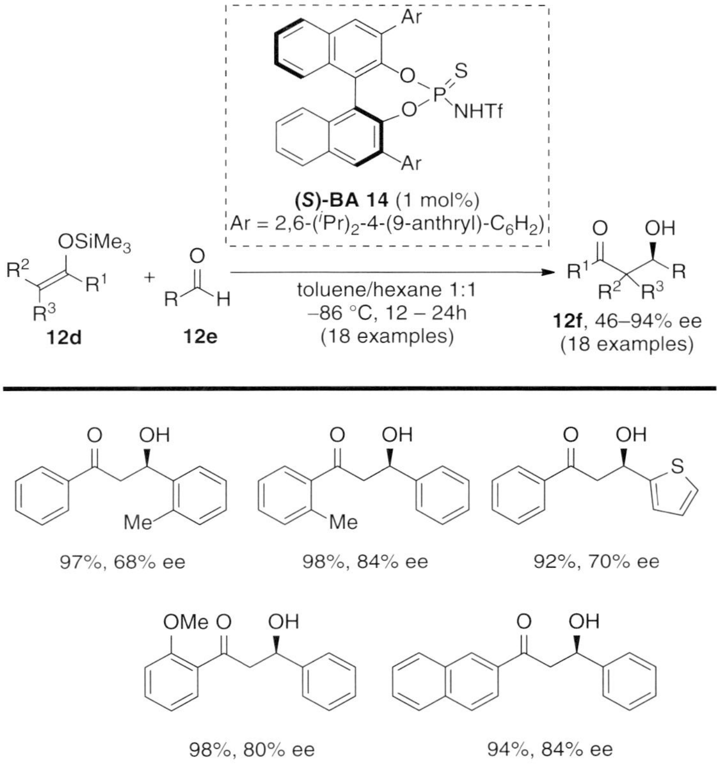

Scheme 4.26 Mukaiyama aldol reaction by Yamamoto [40].

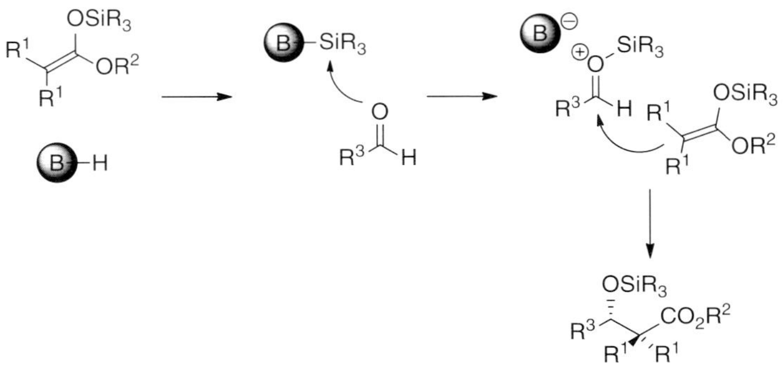

Scheme 4.27 Mechanism of Mukaiyama aldol reaction [38].

silyl group from the silyl ketene acetal to the catalyst and then to the aldehyde. This is followed by a stereoselective addition and regeneration of the silylated catalyst.

Results disclosed by Blanchet in 2010 demonstrated that the Brønsted acidity of the catalyst can in fact be used to control the stereoselectivity to some degree. He showed that coupling of glyoxylates **13a** with generally cyclic ketones **13b** using 5 mol% of **[H$_8$]-PA 5** resulted in the aldol reaction to deliver the products **13c** with moderate enantioselectivities (Scheme 4.28) [41, 42].

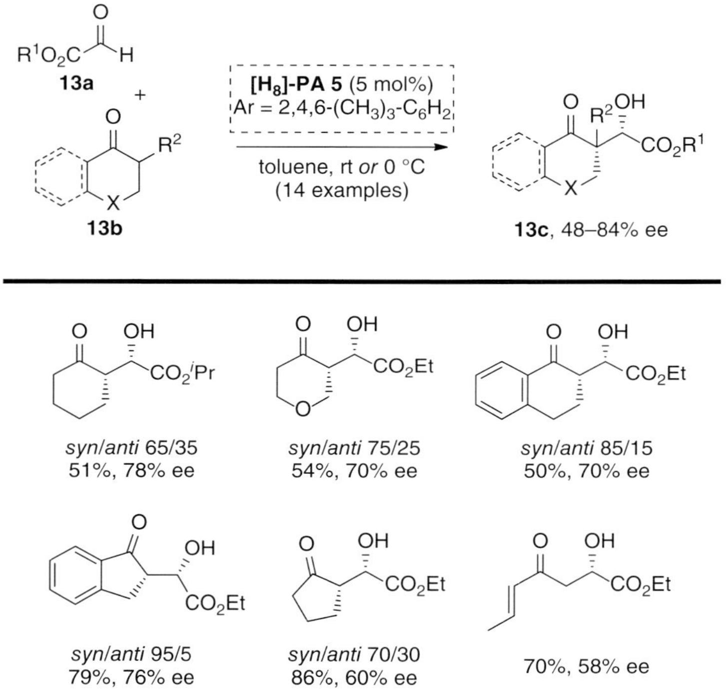

Scheme 4.28 Aldol reaction by Blanchet [41].

The authors noted that although only modest results were achieved, they were superior to conventional covalent enamine catalysis, which is the benchmark for this reaction. In addition, the report also discloses that the catalyst is easily recyclable and performs equally well when reused. A proposed mechanism is depicted in Scheme 4.29. The authors propose a double hydrogen-bonding interaction with the ketone substrate in its enol form and the catalyst. While this may be true, a more conventional view would involve activation of both the enol and the aldehyde substrate during the transition state.

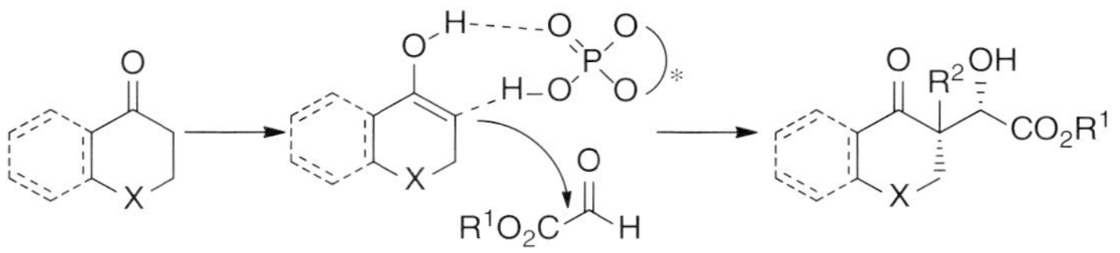

Scheme 4.29 Proposed mechanism of aldol reaction by Blanchet [41].

4.3 Pericyclic Reactions

The hetero Diels–Alder reaction is a useful variant of the classical procedure as it facilitates access to heterocyclic systems, which can be of great value to synthetic chemists. Once again, Brønsted acids have provided great successes albeit in small quantities. In 2009, Terada reported the first example of the title reaction by combining dienes **14a** with glyoxylate **14b** and 5 mol% of **PA 12** to yield dihydropyrans **14c** in exceptional levels of enantioselectivity (Scheme 4.30) [43].

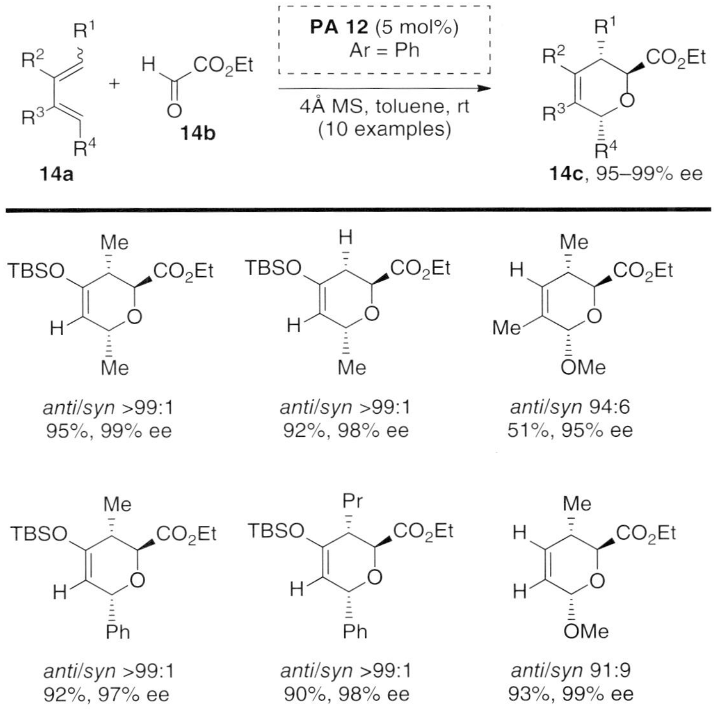

Scheme 4.30 Hetero Diels–Alder by Terada [43].

The reaction proceeds very well with a broad scope of substitution on the diene to deliver *syn*-products. Interestingly, the use of more sterically hindered catalysts provided the anti-products although in lower selectivities. List has also shown a hetero-Diels–Alder reaction to be catalyzed by just 1 mol% of highly fluorinated disulfonimide **BA 15** to access dihydropyrones **14f** with good enantioselectivity (Scheme 4.31) [44]. The reaction boasts an impressive scope; however, mechanistic studies seem to suggest that a stepwise mechanism occurs to deliver the desired products. List has also recently employed a unique disulfonimide catalyst for performing Torgov cyclizations to access tri- and tetra-substituted alkenes [45].

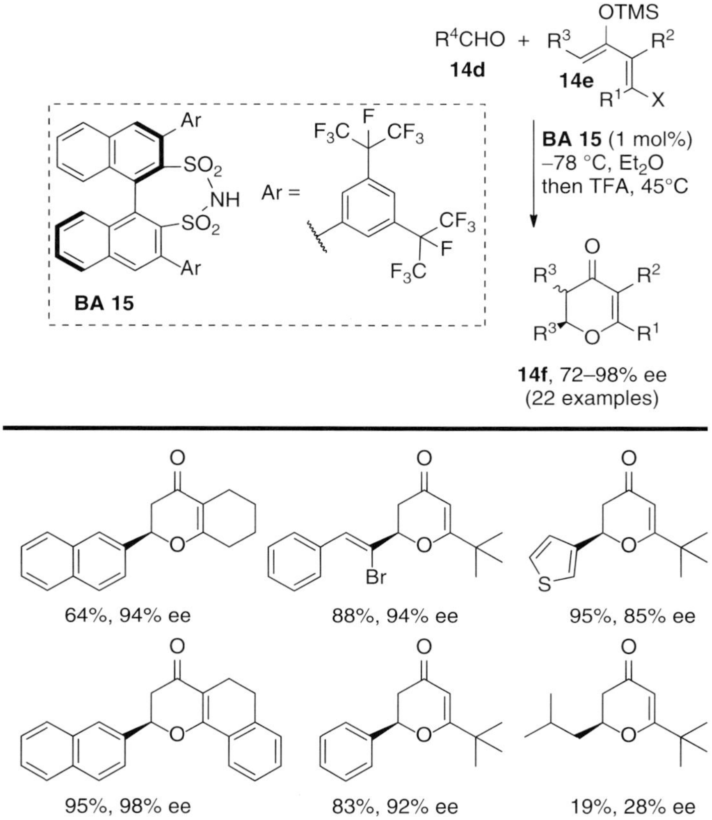

Scheme 4.31 Hetero Diels–Alder by List [44].

The mechanism for the hetero Diels–Alder reaction by Terada is illustrated in Scheme 4.32. It is thought that the catalyst solely activates the glyoxylate aldehyde by a double hydrogen-bonding interaction. This allows the diene to react with the aldehyde and provide the products. The authors propose that for less-hindered catalysts, the diene can occupy the preferred *exo*-orientation for the reaction, which in fact leads to higher selectivities than more sterically hindered catalysts.

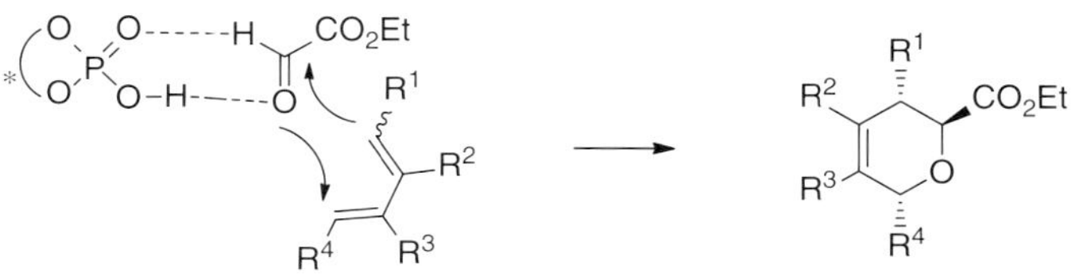

Scheme 4.32 Mechanism of hetero Diels–Alder reaction [43].

The Nazarov cyclization is an electrocyclic process that classically can be found catalyzed by Lewis acids, which coordinate with divinyl ketones to generate a cationic intermediate, which undergoes rearrangement. In 2007, Rueping realized the potential of chiral Brønsted acids to facilitate this process in an asymmetric manner for the first time. Treatment of dienone **15a** with 2 mol% of **NPA 9** resulted in the desired process to yield enantioenriched cyclopentanones **15b** with high enantioselectivities (Scheme 4.33) [46].

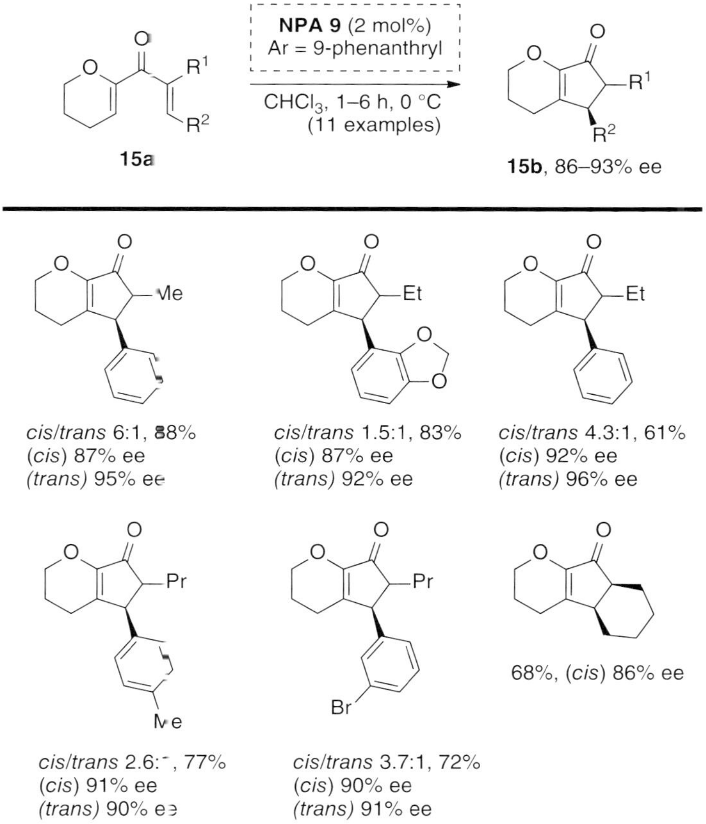

Scheme 4.33 Nazarov cyclizations by Rueping [46].

The reaction at the time constituted the first organocatalyzed electrocyclic reaction and precluded the use of metals or harsh conditions to deliver the products. The groups have gone on to expand the substrate scope and utility of this process in following communications [47–49]. Recently, Tius has also pushed the limits of this transformation to now include vicinal quaternary stereocenters (Scheme 4.34) [50]. He showed that treating **15c** with 10 mol% **NPA 10** resulted in a Nazarov cyclization to yield **15d** generally with excellent enantioselectivities.

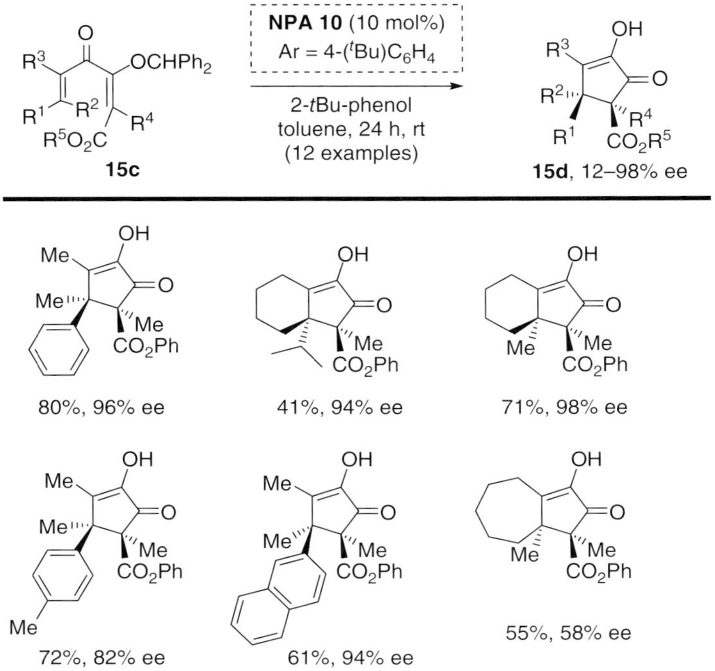

Scheme 4.34 Nazarov cyclizations by Tius [50].

The mechanism of the Nazarov cyclization is illustrated in Scheme 4.35. It is proposed that activation of the carbonyl is facilitated by the catalyst, which is also aided somewhat by the adjacent oxygen atom of the dihydropyran ring. This generates a pentadienyl cation, which undergoes a conrotatory electrocyclization followed by β-hydride elimination to yield an enolate, which can be subsequently protonated.

4.4 Reductions

The asymmetric reduction of ketones to alcohols represents the most efficient process for the generation of chiral secondary alcohols. Although various asymmetric catalytic procedures are known for this process, the use of Brønsted acids

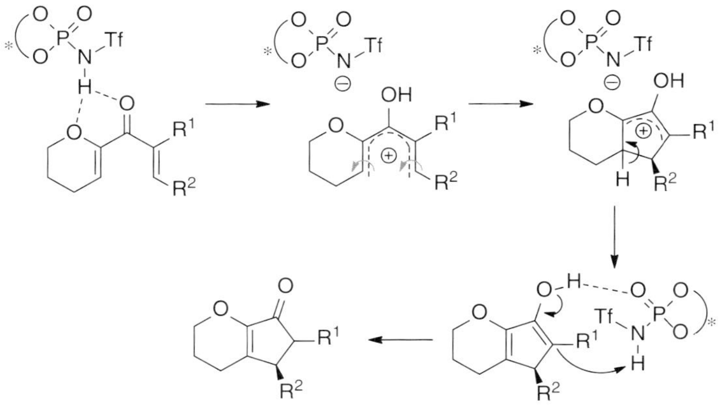

Scheme 4.35 Mechanism of Nazarov cyclization [46].

provides an alternative strategy for synthetic chemists, which may offer access to unique motifs. To date, however, only a single report has been disclosed on this process by the Antilla group. In 2011, he showed that catecholborane **16a** was able to reduce various acetophenones **16b** in the presence of catalytic amounts of **PA 1** and DMAP (Scheme 4.36) [51].

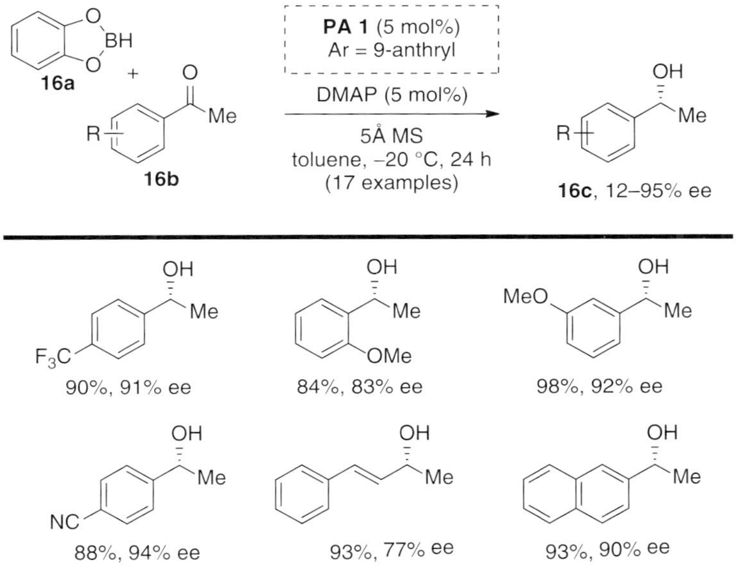

Scheme 4.36 Reduction of ketones by Antilla [51].

In general, the reaction performs well for a range of substrates and provides high to excellent levels of enantioselectivity of **16c**. The proposed mechanism of the process is depicted in Scheme 4.37. The reaction is thought to involve the initial formation of a boron-phosphate species by loss of H_2 gas. It was noted by the authors that gas evolution did occur when the two components were mixed together. This new boron species is then thought to be stabilized by the Lewis basic DMAP additive; however, further studies are needed to verify this hypothesis. The proposed complex then stereoselectively delivers a hydride to the ketone to deliver the products.

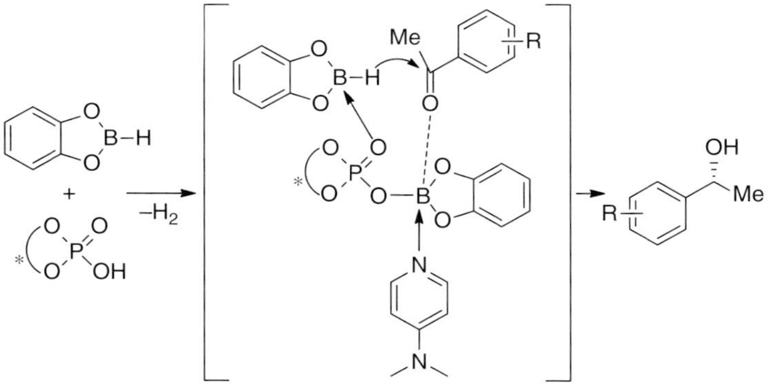

Scheme 4.37 Mechanism of reduction of ketones [51].

References

1. Zeng, X., Ye, K., Lu, M., Chua, P.J., Tan, B., and Zhong, G. (2010) *Org. Lett.*, **12**, 2414–2417.
2. Jain, P. and Antilla, J.C. (2010) *J. Am. Chem. Soc.*, **132**, 11884–11886.
3. Xing, C.-H., Liao, Y.-X., Zhang, Y., Sabarova, D., Bassous, M., and Hu, Q.-S. (2012) *Eur. J. Org. Chem.*, **2012**, 1115–1118.
4. Wang, H., Jain, P., Antilla, J.C., and Houk, K.N. (2013) *J. Org. Chem.*, **78**, 1208–1215.
5. Grayson, M.N., Pellegrinet, S.C., and Goodman, J.M. (2012) *J. Am. Chem. Soc.*, **134**, 2716–2722.
6. Incerti–Pradillos, C.A., Kabeshov, M.A., and Malkov, A.V. (2013) *Angew. Chem. Int. Ed.*, **52**, 5338–5341.
7. Barrio, P., Rodriguez, E., Saito, K., Fustero, S., and Akiyama, T. (2015) *Chem. Commun.*, **51**, 5246–5249.
8. Jain, P., Wang, H., Houk, K.N., and Antilla, J.C. (2012) *Angew. Chem. Int. Ed.*, **51**, 1391–1394.
9. Chen, M. and Roush, W.R. (2012) *J. Am. Chem. Soc.*, **134**, 10947–10952.
10. Tsai, A.S., Chen, M., and Roush, W.R. (2013) *Org. Lett.*, **15**, 1568–1571.
11. Grayson, M.N. and Goodman, J.M. (2013) *J. Am. Chem. Soc.*, **135**, 6142–6148.
12. Reddy, L.R. (2012) *Org. Lett.*, **14**, 1142–1145.
13. Reddy, L.R. (2012) *Chem. Commun.*, **48**, 9189–9191.
14. Rueping, M., Theissmann, T., Kuenkel, A., and Koenigs, R.M. (2008) *Angew. Chem. Int. Ed.*, **47**, 6798–6801.
15. Terada, M., Soga, K., and Momiyama, N. (2008) *Angew. Chem. Int. Ed.*, **47**, 4122–4125.
16. Mahlau, M., García-García, P., and List, B. (2012) *Chem. Eur. J.*, **18**, 16283–16287.
17. Mori, K., Katoh, T., Suzuki, T., Noji, T., Yamanaka, M., and Akiyama, T. (2009) *Angew. Chem. Int. Ed.*, **48**, 9652–9654.

18. Akiyama, T., Katoh, T., and Mori, K. (2009) *Angew. Chem. Int. Ed.*, **48**, 4226–4228.
19. Yamanaka, M., Hoshino, M., Katoh, T., Mori, K., and Akiyama, T. (2012) *Eur. J. Org. Chem.*, **2012**, 4508–4514.
20. Nie, J., Zhang, G.-W., Wang, L., Fu, A., Zheng, Y., and Ma, J.-A. (2009) *Chem. Commun.*, **2009**, 2356–2358.
21. Nie, J., Zhang, G.-W., Wang, L., Zheng, D.-H., Zheng, Y., and Ma, J.-A. (2009) *Eur. J. Org. Chem.*, **2009**, 3145–3149.
22. Kashikura, W., Itoh, J., Mori, K., and Akiyama, T. (2010) *Chem. Asian J.*, **5**, 470–472.
23. Wang, T., Zhang, G.-W., Teng, Y., Nie, J., Zheng, Y., and Ma, J.-A. (2010) *Adv. Synth. Catal.*, **352**, 2773–2777.
24. Yang, M., Zhao, Y.-M., Zhang, S.-Y., Tu, Y.-Q., and Zhang, F.-M. (2011) *Chem. Asian J.*, **6**, 1344–1347.
25. Higuchi, K., Suzuki, S., Ueda, R., Oshima, N., Kobayashi, E., Tayu, M., and Kawasaki, T. (2014) *Org. Lett.*, **17**, 154–157.
26. Wakchaure, V.N. and List, B. (2010) *Angew. Chem. Int. Ed.*, **49**, 4136–4139.
27. Gualtierotti, J.-B., Pasche, D., Wang, Q., and Zhu, J. (2014) *Angew. Chem. Int. Ed.*, **53**, 9926–9930.
28. Harada, S., Kuwano, S., Yamaoka, Y., Yamada, K., and Takasu, K. (2013) *Angew. Chem. Int. Ed.*, **52**, 10227–10230.
29. Mandai, H., Murota, K., Mitsudo, K., and Suga, S. (2012) *Org. Lett.*, **14**, 3486–3489.
30. Qabaja, G., Wilent, J.E., Benavides, A.R., Bullard, G.E., and Petersen, K.S. (2013) *Org. Lett.*, **15**, 1266–1269.
31. Qabaja, G., Benavides, A.R., Liu, S., and Petersen, K.S. (2014) *J. Org. Chem.*, **80**, 133–140.
32. Sano, S., Tsumura, T., Horibe, M., and Nakao, M. (2013) *Synlett*, **24**, 2302–2304.
33. Xu, S., Wang, Z., Zhang, X., Zhang, X., and Ding, K. (2008) *Angew. Chem. Int. Ed.*, **47**, 2840–2843.
34. Xu, S., Wang, Z., Li, Y., Zhang, X., Wang, H., and Ding, K. (2010) *Chem. Eur. J.*, **16**, 3021–3035.
35. Xu, S., Wang, Z., Zhang, X., and Ding, K. (2011) *Eur. J. Org. Chem.*, **2011**, 110–116.
36. Kim, J.H., Čorić, I., Vellalath, S., and List, B. (2013) *Angew. Chem. Int. Ed.*, **52**, 4474–4477.
37. Guin, J., Wang, Q., van Gemmeren, M., and List, B. (2015) *Angew. Chem. Int. Ed.*, **54**, 355–358.
38. García-García, P., Lay, F., García-García, P., Rabalakos, C., and List, B. (2009) *Angew. Chem. Int. Ed.*, **48**, 4363–4366.
39. Ratjen, L., García-García, P., Lay, F., Beck, M.E., and List, B. (2011) *Angew. Chem. Int. Ed.*, **50**, 754–758.
40. Cheon, C.H. and Yamamoto, H. (2010) *Org. Lett.*, **12**, 2476–2479.
41. Pousse, G., Le Cavelier, F., Humphreys, L., Rouden, J., and Blanchet, J. (2010) *Org. Lett.*, **12**, 3582–3585.
42. Das, J., Le Cavelier, F., Rouden, J., and Blanchet, J. (2011) *Eur. J. Org. Chem.*, **2011**, 6628–6631.
43. Momiyama, N., Tabuse, H., and Terada, M. (2009) *J. Am. Chem. Soc.*, **131**, 12882–12883.
44. Guin, J., Rabalakos, C., and List, B. (2012) *Angew. Chem. Int. Ed.*, **51**, 8859–8863.
45. Prévost, S., Dupré, N., Leutzsch, M., Wang, Q., Wakchaure, V., and List, B. (2014) *Angew. Chem. Int. Ed.*, **53**, 8770–8773.
46. Rueping, M., Ieawsuwan, W., Antonchick, A.P., and Nachtsheim, B.J. (2007) *Angew. Chem. Int. Ed.*, **46**, 2097–2100.
47. Rueping, M. and Ieawsuwan, W. (2009) *Adv. Synth. Catal.*, **351**, 78–84.
48. Rueping, M. and Ieawsuwan, W. (2011) *Chem. Commun.*, **47**, 11450–11452.
49. Raja, S., Ieawsuwan, W., Korotkov, V., and Rueping, M. (2012) *Chem. Asian J.*, **7**, 2361–2366.
50. Jolit, A., Walleser, P.M., Yap, G.P.A., and Tius, M.A. (2014) *Angew. Chem. Int. Ed.*, **53**, 6180–6183.
51. Zhang, Z., Jain, P., and Antilla, J.C. (2011) *Angew. Chem. Int. Ed.*, **50**, 10961–10964.

5
Reactions of Generated Carbonyl Intermediates

Unlike imine substrates, carbonyls are generally quite stable structural motifs for usage in organic synthesis. However, the value in developing carbonyl surrogates offers the possibility of unique reactivity modes that are not usually possible from parent carbonyls. For this reason, generated carbonyl intermediates are a growing importance for Brønsted acid catalysis.

5.1
Enol Ethers

From an asymmetric transformation point of view, enol ethers are achiral nucleophilic carbonyl groups that, upon reacting, produce a stereocenter, which may be controlled through the use of a chiral catalyst. Reports of their use with Brønsted acids are generally scarce, but of the methodologies that do exist they provide powerful protocols in organic synthesis. The simplest reaction enol ethers can undergo is protonation, and highly acidic Brønsted acids have shown to be suitable for performing asymmetric protonations. The first report was disclosed by Yamamoto in 2008, who showed that silyl enol ethers **1a** could be asymmetrically protonated by **BA 16** to yield the corresponding α-substituted ketones **1b** with generally high enantioselectivities (Scheme 5.1) [1].

Chiral phosphoric and *N*-triflyl acids failed to facilitate this reaction completely, and so, the increased acidity of **BA 16** was also crucial for this reaction. Recently, Ooi has published the enantioselective protonation of ketene disilyl acetals **1c** using a *P*-chiral diaminodioxaphosphonium barfate catalyst **BA 17** to access α-substituted carboxylic acids **1d** (Scheme 5.2) [2]. In their case, a more sterically hindered phenol was used along with lower temperatures to provide the products with high enantioselectivities. The mechanism of the Yamamoto protonation is given in Scheme 5.3. Combination of the silyl enol ether and the chiral proton source results in protonation and the formation of a carbenium ion. The reaction naturally was highly dependent on the stoichiometric achiral proton source, and in their case, they found that phenol provided the most optimal results, which also included the removal of the trimethylsilyl (TMS) group.

Asymmetric Brønsted Acid Catalysis, First Edition. Magnus Rueping, Dixit Parmar and Erli Sugiono.

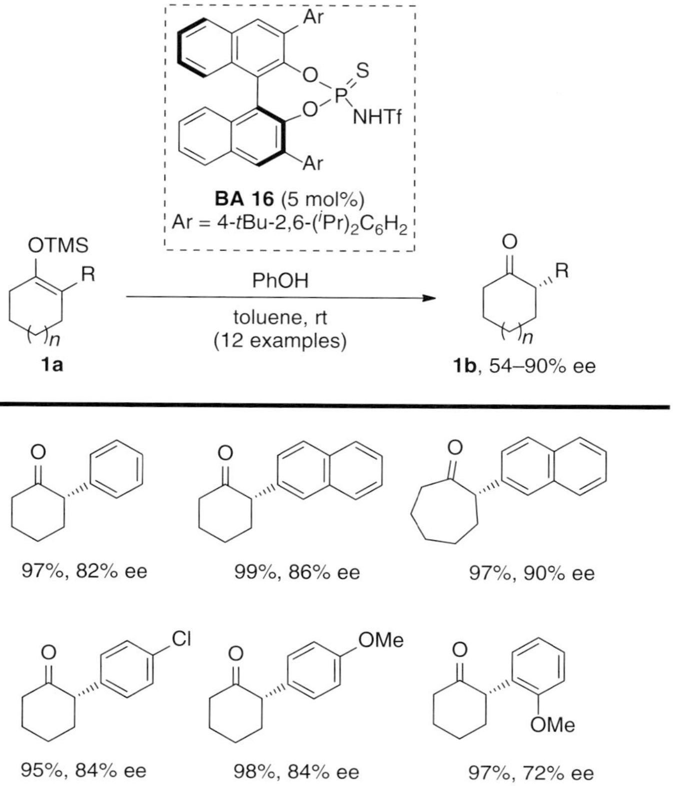

Scheme 5.1 Enantioselective protonations by Yamamoto [1].

Another feature of enol ethers under acidic conditions is that upon protonation, they can generate a highly electrophilic oxonium species, which contain bound chiral counterions (derived from the catalyst) that not only improve their inherently poor reactivity but also facilitate the enantioselective addition of nucleophiles to the carbonyl carbon atom. This feature was exploited for the first time by List in 2012, who demonstrated the asymmetric spiroacetalization reaction of hydroxyenol ethers **2a** with C_2-symmetric imidodiphosphoric acid catalyst **NPA 11** (Scheme 5.4) [3].

The report is focused on the development of C_2-symmetric imidodiphosphoric catalysts as advancements to the more routinely utilized phosphoric acid catalysts. From the analysis of crystal structures, this new blend of catalysts offer a much smaller and more sterically constrained site of reactivity, which can be considered as some progress toward an enzyme-like pocket that is found in nature. The chiral acetal products **2b** are obtained with excellent enantioselectivities especially considering the absence of any additional structural motifs.

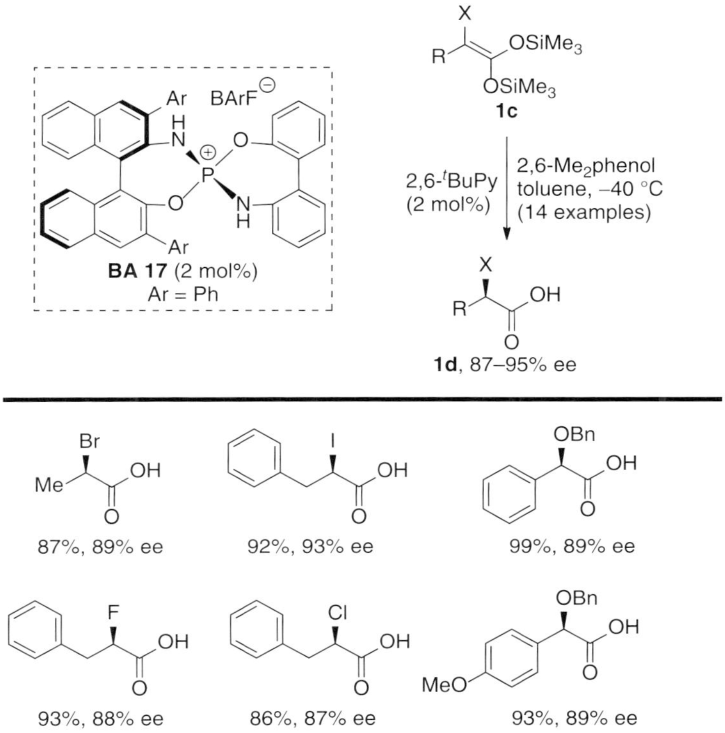

Scheme 5.2 Enantioselective protonations by Ooi [2].

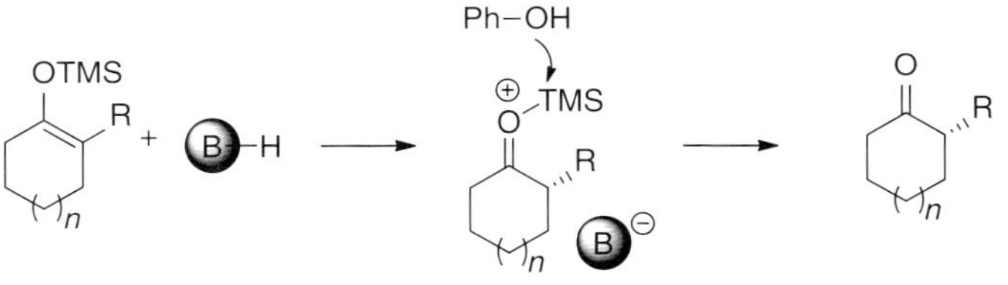

Scheme 5.3 Mechanism of protonations [1].

The mechanism of the reaction is shown in Scheme 5.5. It is proposed that the catalyst is able to activate both the electrophilic and nucleophilic sites of the substrates. This initially leads to a protonation of the cyclic enol ether to generate an oxocarbenium ion, which can be intercepted by the pendant alcohol group. In the same year, Nagorny also showed an enantioselective and diastereoselective spiroacetalization reaction that could be catalyzed by a chiral phosphoric acid, which delivered good to high results [4].

Enol ethers have also found some utility as electrophilic species toward asymmetric intermolecular reactions. For example, an isolated report by Floreancig has shown the addition of nitriles to vinyl ethers [5]. In 2009, Terada reported a direct

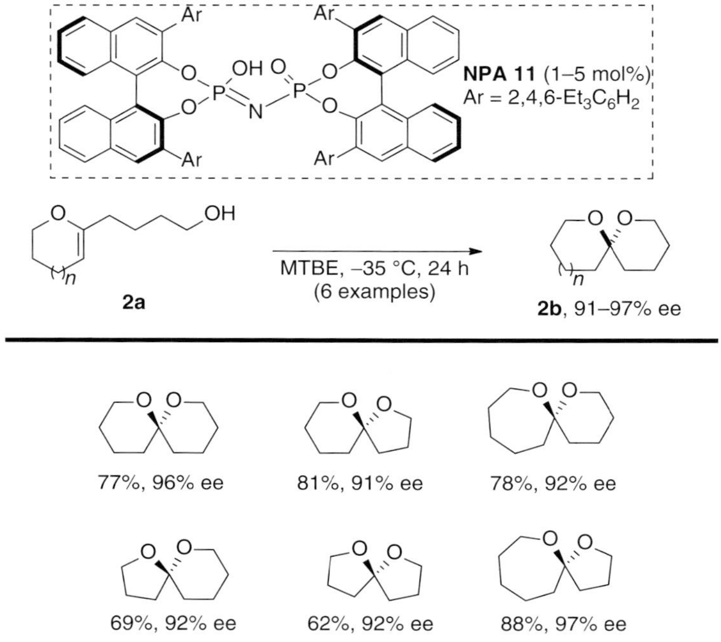

Scheme 5.4 Spiroacetalization using a C_2-symmetric catalyst by List [3].

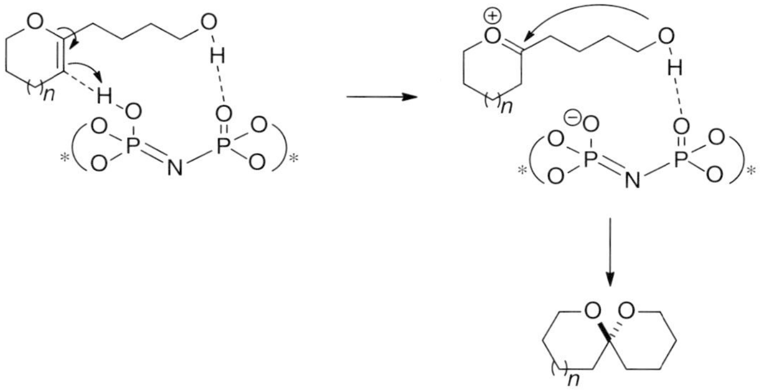

Scheme 5.5 Mechanism of spiroacetalization [3].

aldol-type reaction of vinyl ethers **3a** with azlactones **3b** in the presence of 5 mol% **PA 16** (Scheme 5.6) [6].

The proposed mechanism of the reaction is portrayed in Scheme 5.7. Initially, protonation of the starting material by the catalyst generates an oxocarbenium ion, which is inherently paired with the phosphate anion. The authors suggest that a number of hydrogen bonds may be present within this intermediate. The enol tautomer of the azlactone is then able to attack this intermediate and form an

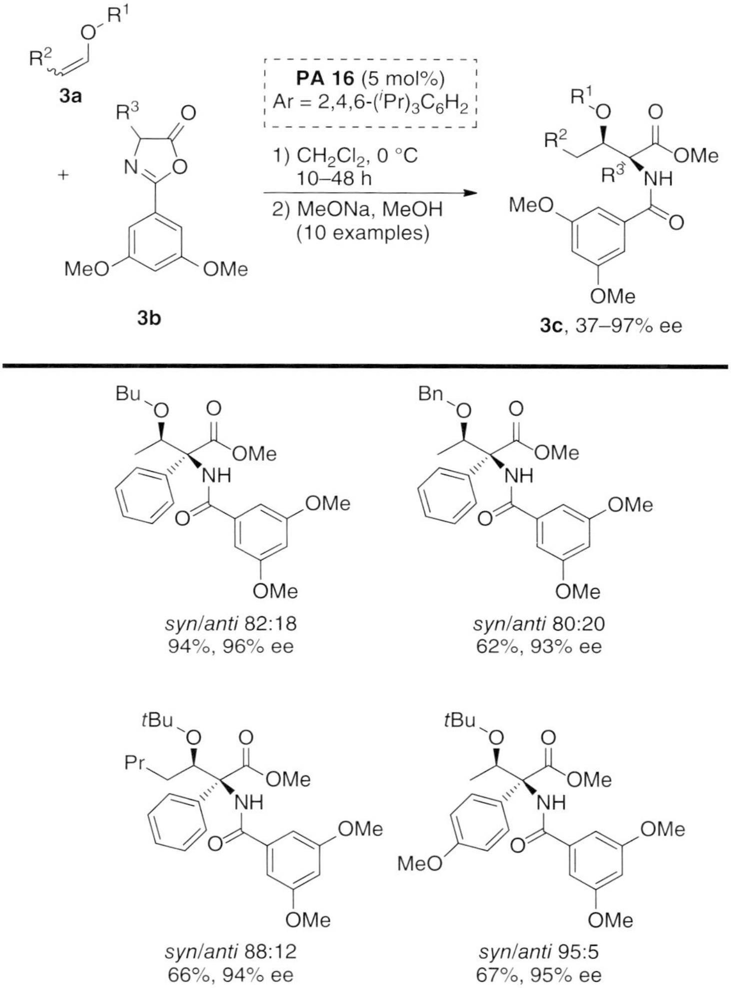

Scheme 5.6 Aldol-type reaction by Terada [6].

adduct, which can be hydrolyzed under basic conditions to deliver β-hydroxy-α-amino acid derivatives in generally high enantioselectivities.

5.2
Acetals

Acetals share a common trait with enol ethers in that under acidic conditions, they too can generate oxocarbenium ions, which potentially can contain the chiral

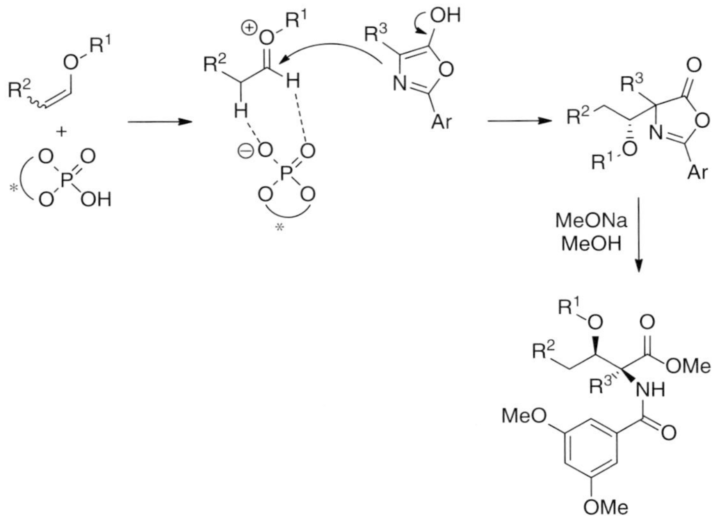

Scheme 5.7 Mechanism of aldol-type reaction [6].

counterion of a catalyst to promote stereoselective reactions. List recognized this potential in 2010, wherein he showed that hydroxyl-acetals **4a** could be treated with **(*S*)-PA 16** to undergo a cyclative transacetalization process to yield acetals **4b** with good selectivities (Scheme 5.8) [7, 8].

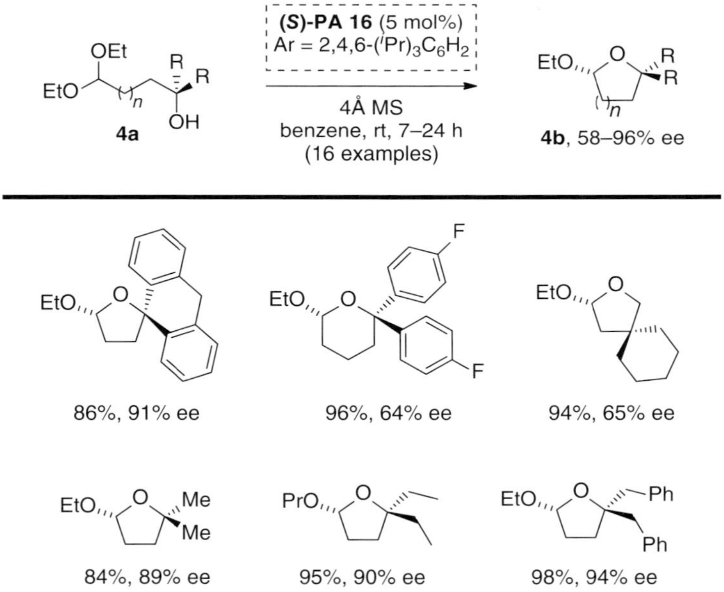

Scheme 5.8 Transacetalization by List [7, 8].

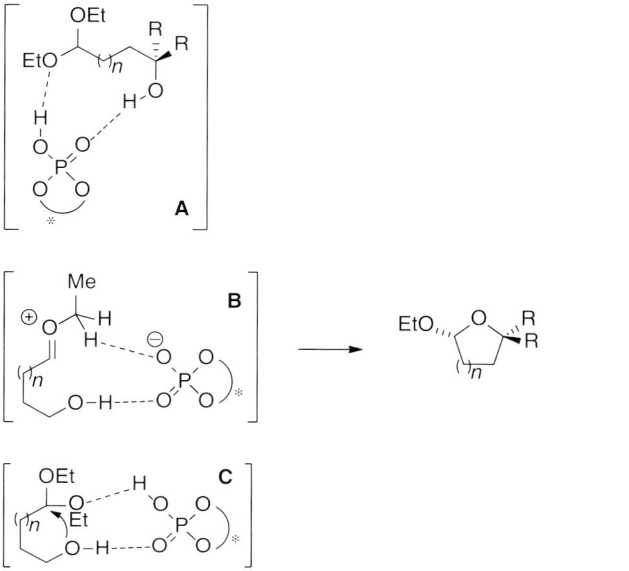

Scheme 5.9 Proposed transition states for transacetalization [7, 8].

With regard to the mechanism, a number of possibilities are feasible, which account for the reactivity observed (Scheme 5.9). Pathway A could involve a hydrogen-bonding network, which involves the selective activation of one acetal group. This is, however, not considered to be occurring. The two dominant pathways proposed involve either acetal collapse to reveal an oxocarbenium ion, which is trapped by the hydroxyl group or a S_N2-like pathway involving direct attack. In addition, the phosphoric acid catalyst may be involved in multiple coordinations with the substrate during the reaction process. Sun has also shown a similar procedure in a desymmetrization process to access tetrahydrofurans and tetrahydropyrans [9].

The combination of photo-mediated transformations with chiral Brønsted acid catalysis is a novel concept that has been employed recently by the Rueping group in performing asymmetric hydrogenations of *in situ* formed pyrylium ions. The group demonstrated that phenolic enones **5a** could undergo a light-mediated cyclization to form unstable chroman-2-ols **5b,** which upon loss of H_2O can be subsequently hydrogenated with Hantzsch ester to yield the products **5c** with good enantioselectivities (Scheme 5.10) [10].

The mixture of light and **[H$_8$]-PA 16** was shown to have a synergistic effect on the reaction as both components were shown to be required throughout the complete transformation. In the same year, Terada published a related procedure starting directly from chroman-2-ols **5d** to access the hydrogenated products **5e** with good selectivities using **PA 16** (Scheme 5.11) [11].

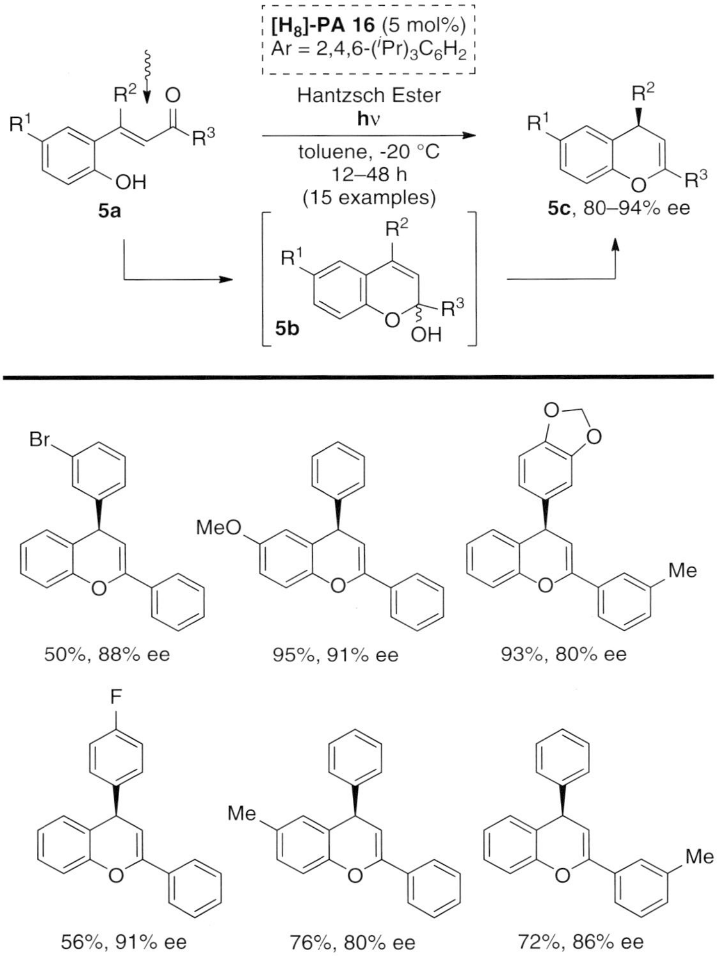

Scheme 5.10 Hydrogenation of pyrylium ions by Rueping [10].

A rather interesting use of acetals was published recently by Houk and Zheng regarding a desymmetrization oxidation reaction. Treating acetals **6a** with DMDO under the influence of 5 mol% **(*S*)-PA 16** resulted in a smooth asymmetric oxidation to yield the chiral alcohols **6b** with good enantioselectivities (Scheme 5.12) [12].

Formally, the process can be thought of as a desymmetrization of diols, since the starting acetals **6a** can be prepared from the corresponding diols. The group also performed DFT studies, which revealed that the aryl–aryl interactions between the substrate and the catalyst are crucial to achieving high enantioselectivites.

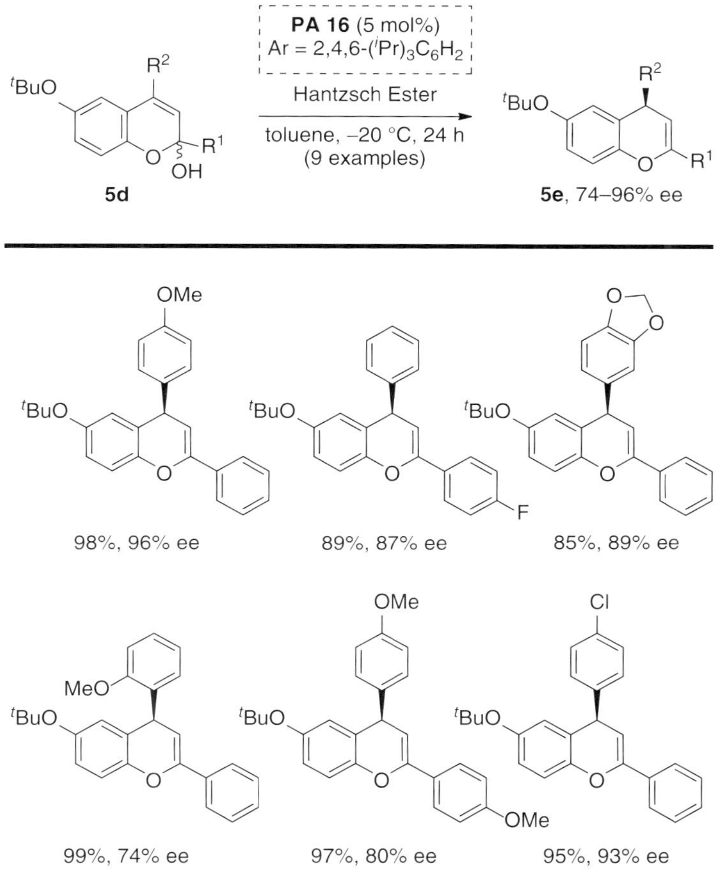

Scheme 5.11 Hydrogenation of pyrylium ions by Terada [11].

5.3 Phenols Containing Leaving Groups

This section covers aromatic alcohol or ether substrates, which formally contain a leaving group, usually OH. The group is suitably aligned for elimination and thus forming an electrophilic quinine intermediate (also viewed as a pseudo-oxocarbenium ion), which can be activated by the chiral anion of the Brønsted acid catalyst. Although in most cases, the reaction does not function in the absence of these donating groups, we cannot assume for every case that this is the sole pathway of the process. This section bears close similarities with Section 3.2 on indoles containing leaving groups, which behave as nitrogen equivalents.

Substrates in this category regardless of the mechanism possess a certain electrophilic character; however, only recently have research groups been able

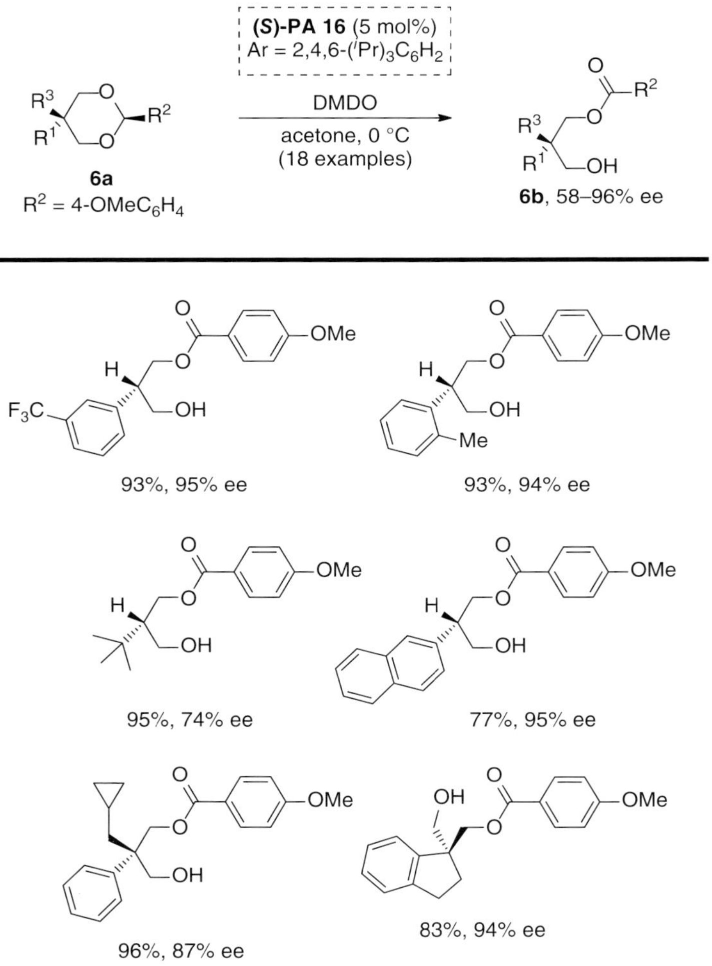

Scheme 5.12 Desymmetrization of diols by Zheng [12].

to control this process with acceptable levels of selectivity. An early example of this was shown by Bach, who performed the addition of indoles to *ortho*-hydroxybenzylic alcohols, which gave low to modest enantioselectivities (Scheme 5.13) [13].

Treating **7a** with indole **7b** in the presence of **[H$_8$]-PA 12** gave the corresponding addition product **7c** in a modest 77% ee. The mechanism of the process is intriguing and could follow a number of possible pathways (Scheme 5.14). The authors propose that a discreet carbocation is formed upon mixing the substrate with the phosphoric acid catalyst. While this may be occurring, the adjacent aromatic group also contains electron-donating groups. They can also participate in

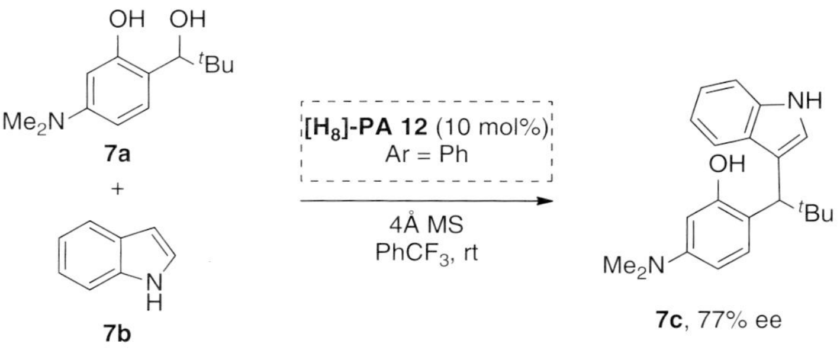

Scheme 5.13 Addition of indoles to *ortho*-hydroxybenzylic alcohols by Bach [13].

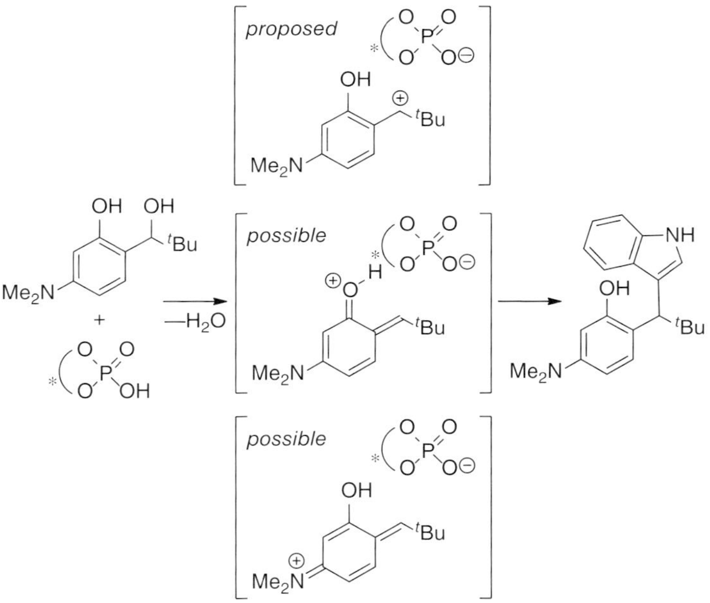

Scheme 5.14 Mechanism of indole addition to *ortho*-hydroxybenzylic alcohols [13].

the reaction mechanism by the donation of electrons to eliminate H_2O. Once the intermediate cationic species is formed, it can be captured by the indole to yield the observed products.

In 2014, the groups of Rueping and Schneider have showed that 1,3-dicarbonyls can be used as suitable nucleophiles for intercepting these electrophilic intermediates. Rueping reported that **8b** could be used as a coupling partner for 2-hydroxy benzylic alcohols **8a** in the presence of 5 mol% **[H$_8$]-PA 22** to yield tetrahydroxanthenes **8c** with high enantioselectivities (Scheme 5.15) [14].

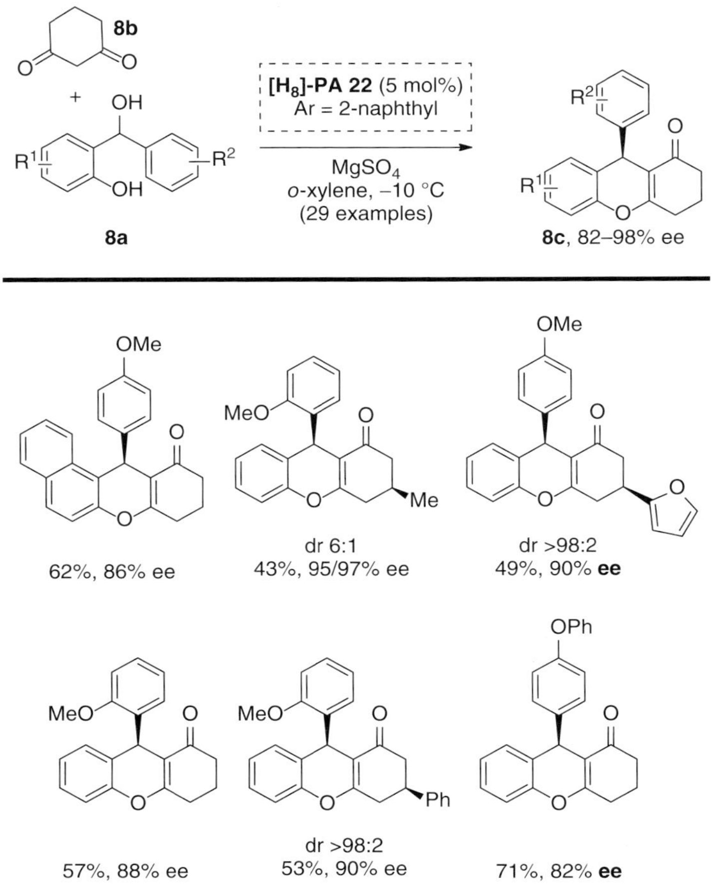

Scheme 5.15 Synthesis of substituted tetrahydroxanthene by Rueping [14].

Schneider chose to employ sterically hindered catalyst **PA 34,** which was able to deliver the products **8f** after an additional TsOH-mediated condensation step in similar enantioselectivities (Scheme 5.16) [15]. In addition, Schneider and Sun have also shown that indoles and 2-naphthols are suitable nucleophiles for this process [16, 17]. The mechanism of the reaction is given in Scheme 5.17. It is proposed that under acidic conditions, elimination of H_2O from the substrate results in an *ortho*-quinone intermediate. This contains a bound phosphate ion, which is hydrogen-bonded to the phenol proton. That allows for the enol form of the 1,3-dicarbonyl to attack, resulting in the products observed.

A rather unique example has been recently disclosed by Hong, who has shown the kinetic resolution of racemic diols for the application of total synthesis. By taking racemic **(±)-9a** in the presence of 14 mol% of **(*S*)-PA 16**, the undesired

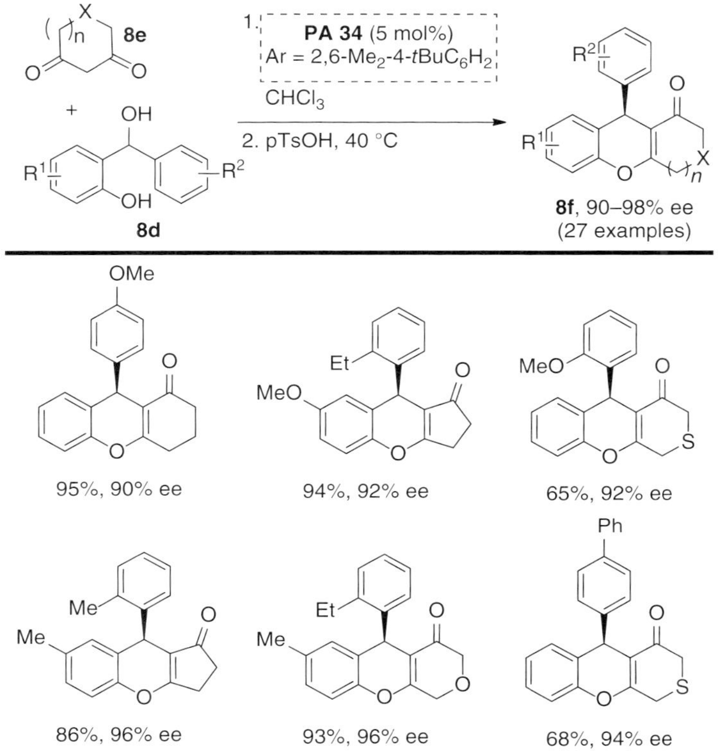

Scheme 5.16 Synthesis of substituted tetrahydroxanthene by Schneider [15].

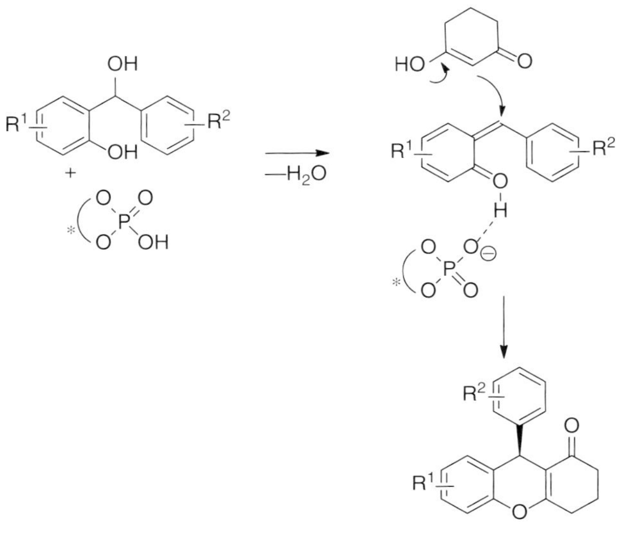

Scheme 5.17 Mechanism of *ortho*-quinone methide reactions [14].

enantiomer could be selectively removed by a catalyst promoted cyclization leaving behind **(+)-9a** in an excellent 97% ee (Scheme 5.18) [18].

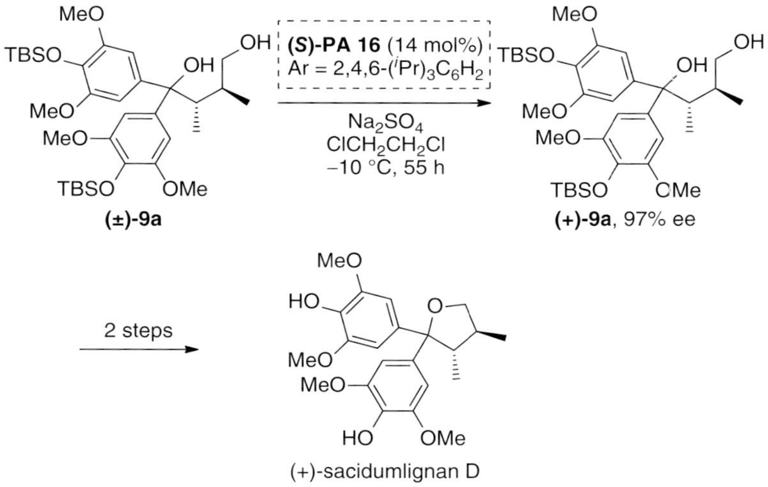

Scheme 5.18 Total synthesis (+)-sacidumlignan D by Hong [18].

The group also noticed that the addition of Na_2SO_4 was beneficial to the kinetic resolution process. This was proposed to be responsible for capturing H_2O molecules released from the undesired enantiomer. Just two further straightforward steps were then required to complete the total synthesis of (+)-sacidumlignan D.

References

1. Cheon, C.H. and Yamamoto, H. (2008) *J. Am. Chem. Soc.*, **130**, 9246–9247.
2. Uraguchi, D., Kizu, T., Ohira, Y., and Ooi, T. (2014) *Chem. Commun.*, **50**, 13489–13491.
3. Čorić, I. and List, B. (2012) *Nature*, **483**, 315–319.
4. Sun, Z., Winschel, G.A., Borovika, A., and Nagorny, P. (2012) *J. Am. Chem. Soc.*, **134**, 8074–8077.
5. Lu, C., Su, X., and Floreancig, P.E. (2013) *J. Org. Chem.*, **78**, 9366–9376.
6. Terada, M., Tanaka, H., and Sorimachi, K. (2009) *J. Am. Chem. Soc.*, **131**, 3430–3431.
7. Čorić, I., Vellalath, S., and List, B. (2010) *J. Am. Chem. Soc.*, **132**, 8536–8537.
8. Čorić, I., Müller, S., and List, B. (2010) *J. Am. Chem. Soc.*, **132**, 17370–17373.
9. Chen, Z. and Sun, J. (2013) *Angew. Chem. Int. Ed.*, **52**, 13593–13596.
10. Hsiao, C.-C., Liao, H.-H., Sugiono, E., Atodiresei, I., and Rueping, M. (2013) *Chem. Eur. J.*, **19**, 9775–9779.
11. Terada, M., Yamanaka, T., and Toda, Y. (2013) *Chem. Eur. J.*, **19**, 13658–13662.
12. Meng, S.-S., Liang, Y., Cao, K.-S., Zou, L., Lin, X.-B., Yang, H., Houk, K.N., and Zheng, W.-H. (2014) *J. Am. Chem. Soc.*, **136**, 12249–12252.
13. Wilcke, D., Herdtweck, E., and Bach, T. (2011) *Synlett*, (9), 1235–1238.
14. Hsiao, C.-C., Liao, H.-H., and Rueping, M. (2014) *Angew. Chem. Int. Ed.*, **53**, 13258–13263.

15. El-Sepelgy, O., Haseloff, S., Alamsetti, S.K., and Schneider, C. (2014) *Angew. Chem. Int. Ed.*, **53**, 7923–7927.

16. Saha, S., Alamsetti, S.K., and Schneider, C. (2015) *Chem. Commun.*, **51**, 1461–1464.

17. Zhao, W., Wang, Z., Chu, B., and Sun, J. (2015) *Angew. Chem. Int. Ed.*, **54**, 1910–1913.

18. Xie, C., Bai, D., Huang, S.-H., Jia, X., and Hong, R. (2014) *Asian J. Org. Chem.*, **3**, 277–280.

6 Reactions of Alkenes

The activation of alkenes by chiral Brønsted acid catalysts is regarded as a major area of research alongside imine substrates. Since alkenes are versatile substrates, amenable to a plethora of reactions, they represent an important class of substrates for Brønsted acid catalysis. In general, however, isolated alkenes lacking any stabilizing groups are incredibly difficult substrates to activate and for that reason their reports in asymmetric transformations are extremely scarce. Alkenes possessing electron-withdrawing substituents (e.g., esters) are more readily activated. Although formally the following reports may be activated by hydrogen-bonding interactions with groups adjacent to the alkene, reactivity occurs at the alkene and therefore they have been included in this section.

6.1 Nucleophilic Addition Reactions

The nucleophilic addition reaction is a broad topic that contains various valuable transformations depending on the nature of the nucleophile used. Surprisingly, simple C-nucleophiles are not present in this section; however, alkyl boronate additions have been demonstrated with dicarboxylic acid catalysts [1, 2]. The addition of nitrogen and oxygen nucleophiles to alkenes is a moderately studied area of alkene-activated reactions and contains a number of highly useful methodologies. As alluded to in the introduction to this section, alkenes containing no stabilizing groups are difficult substrates to activate and as such reports of their use are scarce. In 2008, a report from Ackermann demonstrated with modest results the first example of such activation. He showed that unsaturated amine **1a** in the presence of 20 mol% **PA 13** gave the cyclized pyrrolidine **1b** in 17% ee (Scheme 6.1, top) [3].

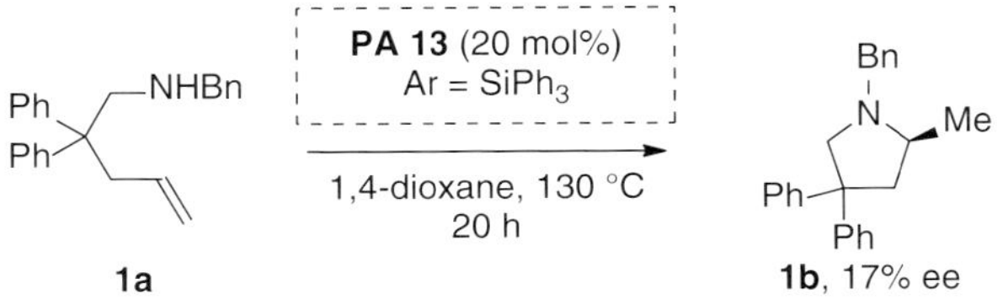

Scheme 6.1 Hydroamination by Ackermann [3].

Asymmetric Brønsted Acid Catalysis, First Edition. Magnus Rueping, Dixit Parmar and Erli Sugiono.

The mechanism is illustrated in Scheme 6.2. Although it is not proposed by the authors, the most likely pathway may involve coordination of the N–H proton to the catalyst with simultaneous activation of the alkene to yield the desired products.

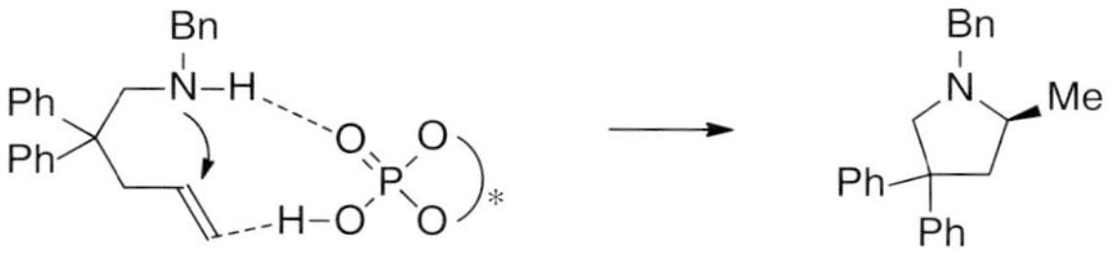

Scheme 6.2 Mechanism of hydroamination [3].

In 2011, Toste devised a dithiophosphoric catalyst (**BA 18**), which could overcome this deficiency with impressive results. By taking amino-dienes **2a** as substrates, he showed that they could be cyclized to yield a variety of five-membered N-heterocycles **2b** with excellent levels of control (Scheme 6.3) [4].

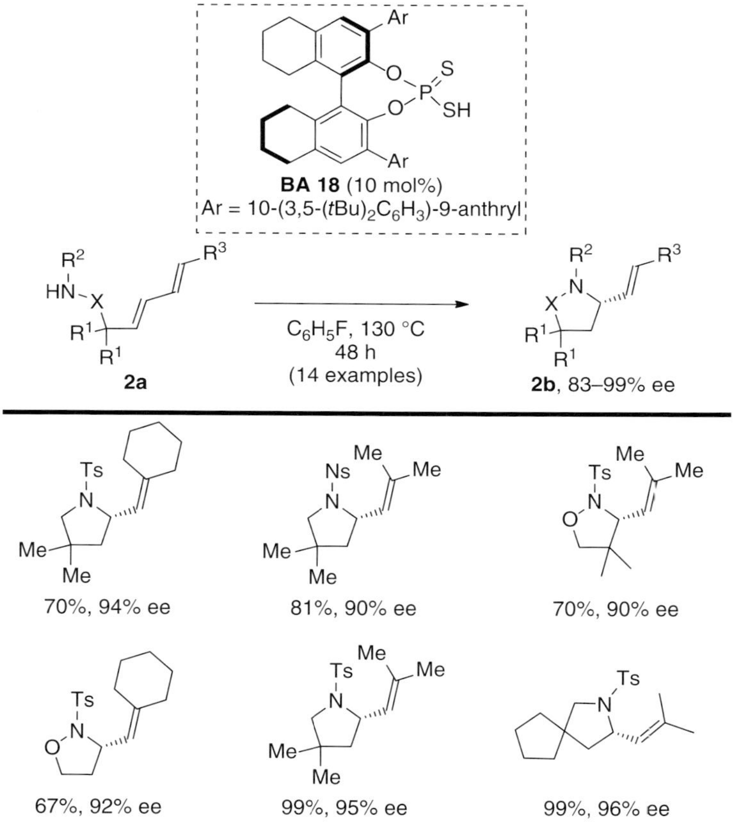

Scheme 6.3 Hydroamination by Toste [4].

The mechanism of the reaction is proposed in Scheme 6.4. Mechanistic studies revealed that the origin of selectivity arises from the initial hydrothiolation of the distal alkene unit by the catalyst followed by subsequent displacement by the amine group to release the catalyst. Zimmerman and Nagorny have also recently studied a similar transformation of α,β-unsaturated acetals [5].

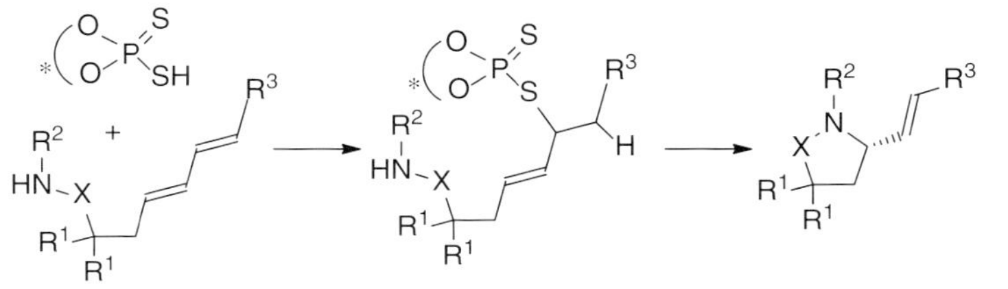

Scheme 6.4 Mechanism of hydroamination [4].

The Ooi group has been pioneers of the use of phosphonium barfates as highly efficient Brønsted acid catalysts. For example, they have shown that just 2 mol% of catalyst **BA 19** can catalyze the coupling of nitroalkenes **3a** with aniline **3b** to give the adducts **3c** with excellent enantioselectivities (Scheme 6.5) [6, 7].

During the optimization process, it was noted that the absolute configuration of the individual binaphthyl units played a crucial role in the level of selectivity observed. **BA 19** possesses an (*R*,*S*)-unit respectively (from left to right); however, the corresponding (*R*,*R*)-isomer gave considerably lower levels of control.

A rather unique example of the activation of quinone imine ketals was presented by Maruoka in 2013 using an axially chiral dicarboxylic acid catalyst (**BA 1**). It was shown that quinone imine ketals **4a** could be coupled with enecarbamates **4b** to give the re-aromatized α-amino-β-aryl ethers **4c** with high enantioselectivities (Scheme 6.6) [8].

Generally, the reaction proceeds smoothly for a range of substitutions for the various R-groups; however, the use of a carbamate for the quinone imine ketals-protecting group gave optimal results. The products themselves were shown to be highly useful synthetic fragments for a range of additional manipulations.

The mechanism of the reaction is illustrated in Scheme 6.7. Activation of the quinone imine ketal occurs by the acid, which then allows the enecarbamate to attack to generate an iminium intermediate. This substrate can then rearomatize by loss of MeOH, which can come back to attack the iminium ion and thus yield the products. The catalyst may also be involved in coordinating and activating the enecarbamate substrate. Recently, Wang and Shi have shown styrenes to be suitable nucleophiles for quinone imine ketals activated by a phosphoric acid catalyst [9].

α,β-Unsaturated carbonyl groups are highly suited for activation as the carbonyl both provides a point of contact for the catalyst and lowers the LUMO energy of the alkene, making it suitable for reactivity. In 2010, You reported on an aza-Michael reaction of indoles **5a** using **(*S*)-PA 13** in 10 mol% to access the polycyclic products **5b** generally with good enantioselectivities (Scheme 6.8) [10].

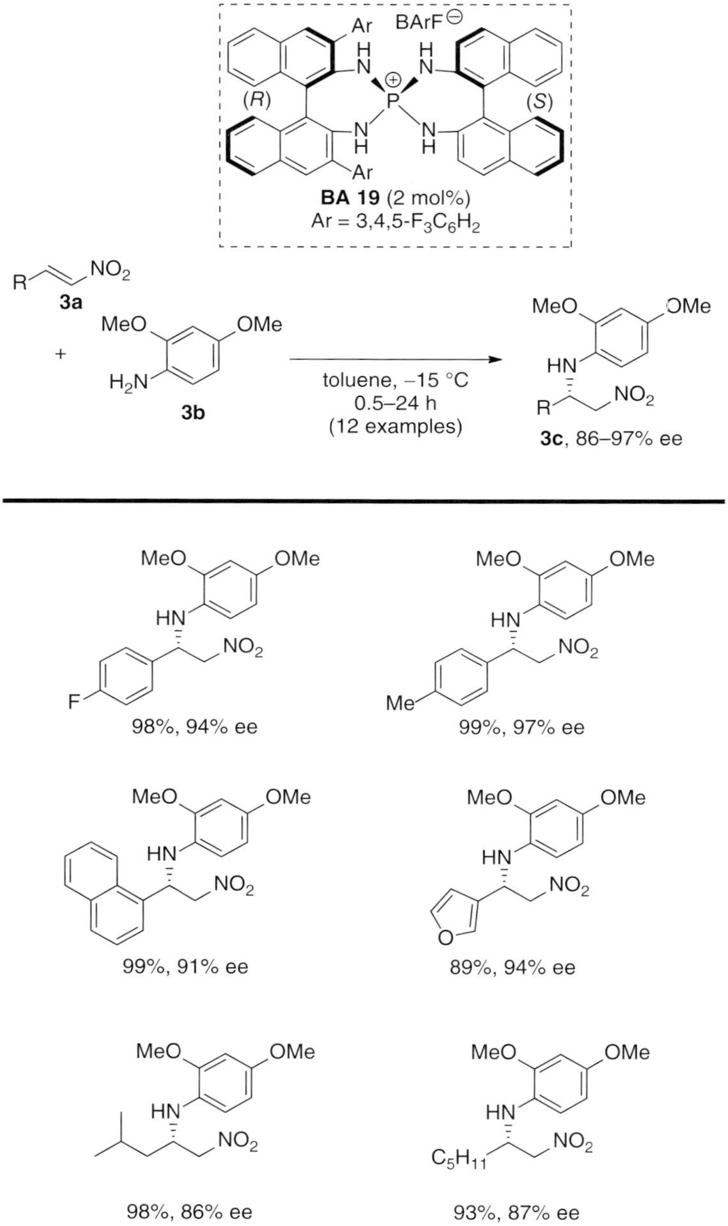

Scheme 6.5 Addition of arylamines to nitroolefins by Ooi [6].

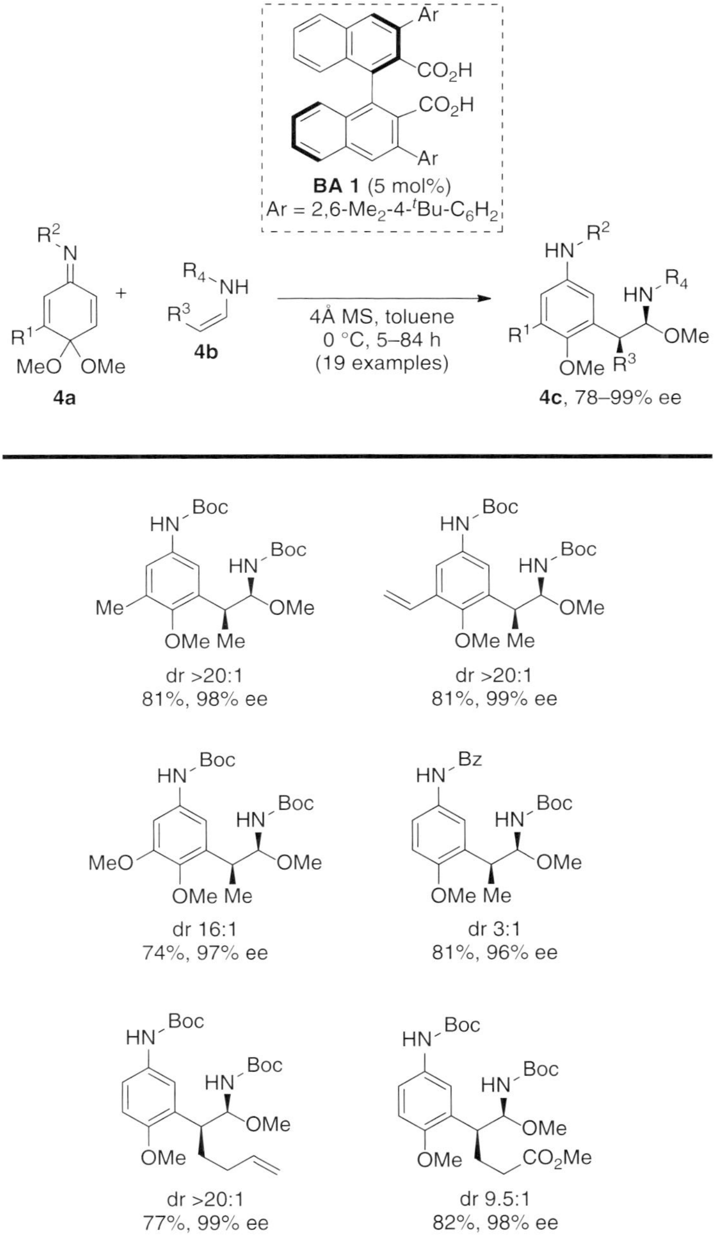

Scheme 6.6 Activation of quinone imine ketals by Maruoka [8].

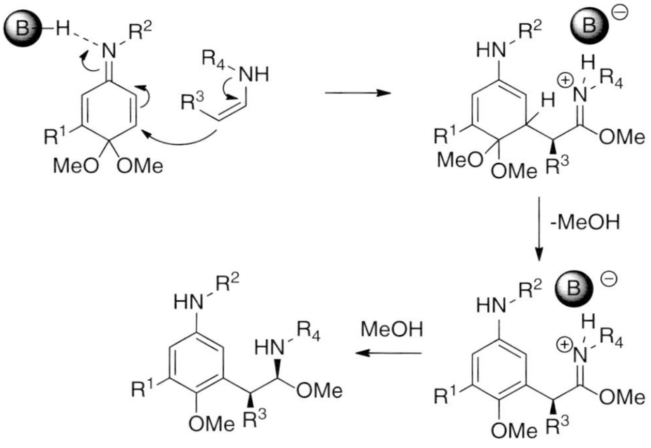

Scheme 6.7 Activation of quinone imine ketals.

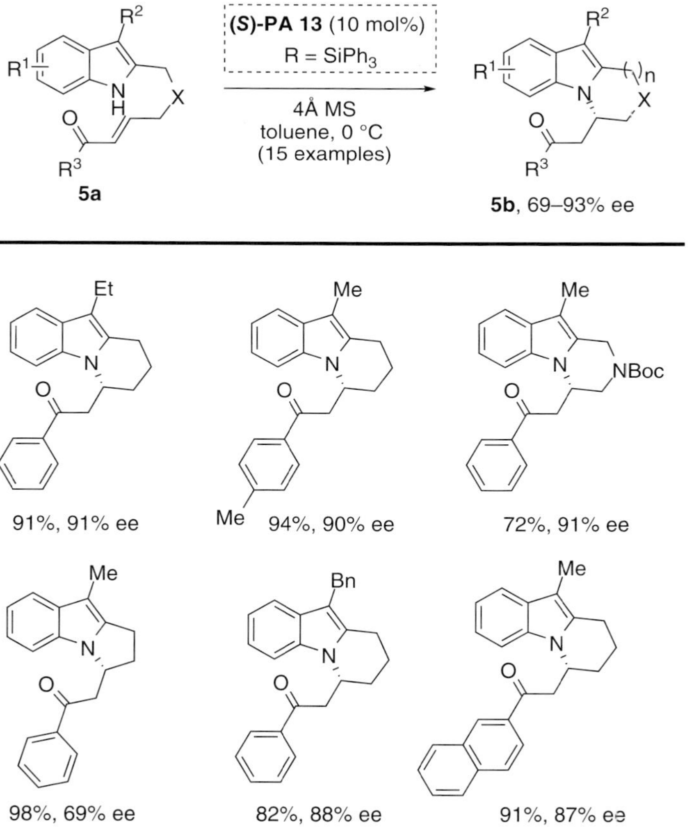

Scheme 6.8 Aza-Michael reaction of indoles by You [10].

The reaction performs smoothly for a range of substrates and ring sizes in good overall yields. It was noted that aromatic substitution only delivered modest selectivities; however, the use of a bulky silicon substituent provided the optimal results. In addition, the use of 4 Å molecular sieves also produced a faster rate and better enantioselectivities. You has also developed an aza-Michael reaction to access dihydroquinolines [11] and an oxo-Michael reaction [12]. Closely related studies on the intramolecular aza-Michael reaction have also been conducted by other research groups [13, 14].

Oxygen nucleophiles are less well studied than their nitrogen counterparts for this section, which is slightly surprising given that nitrogen nucleophiles usually need appropriate protecting groups to function well. One of the first reports was by Rueping in 2011, who showed that 2-substituted phenols **6a** when treated with **[H_8]-NPA 12** underwent cyclative allylic substitutions to yield 2H-chromenes **6b** with high enantioselectivites (Scheme 6.9) [15].

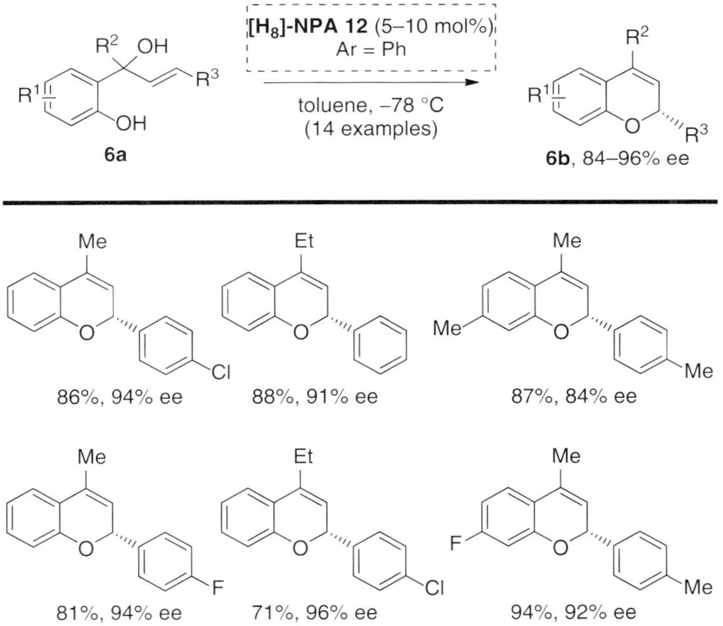

Scheme 6.9 Contact ion pair allylic substitution by Rueping [15].

The mechanism of the reaction is depicted in Scheme 6.10. It is thought that firstly a loss of H_2O occurs, which sets up a chiral contact ion pairing between the catalyst and the allylic cation generated. This intermediate then proceeds to be intercepted by the phenol leading to the products. Recently, Du has used a similar system to perform intermolecular aminations [16] and Gong has used the system in total synthesis [17].

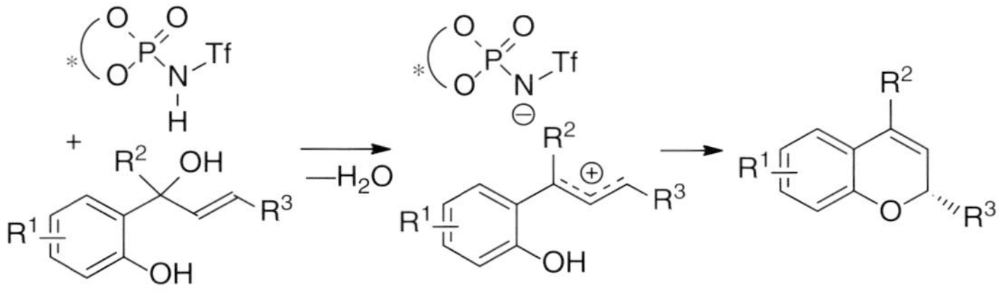

Scheme 6.10 Mechanism of contact ion pair allylic substitution [15].

A rather unique catalyst motif consisting of a chiral phosphorus catalyst has been shown in an isolated example by Ishihara to be capable of catalyzing a kinetic resolution process with moderate results. Treating **(±)-7a** with 20 mol% of **BA 20** resulted in selective cyclization of a single enantiomer to form lactones **7b** (Scheme 6.11) [18].

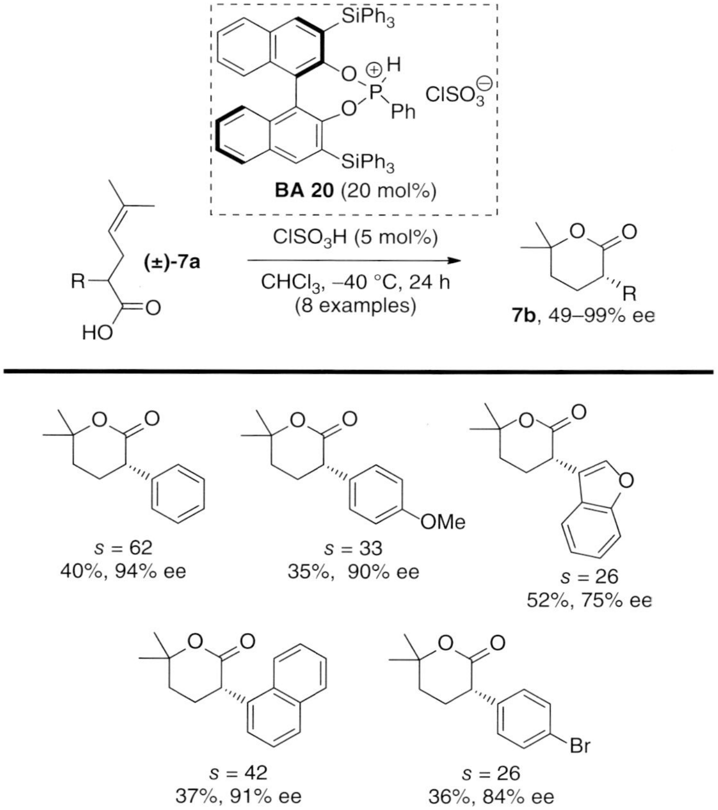

Scheme 6.11 Kinetic resolution of carboxylic acids by Ishihara [18].

A catalytic quantity of $ClSO_3H$ was also required as a co-acid to regenerate the catalytic species, which was chosen since it did not interfere with the enantioselectivity of the reaction. The unreacted carboxylic acid enantiomer could also be recovered in high enantiomeric excess.

6.2 Friedel–Crafts Reactions

The Friedel–Crafts reaction is a highly studied reaction by research groups as Brønsted acids have proved to be highly suitable for the activation of substrates, which can be coupled with aromatic partners. In that respect, alkenes have managed to provide a significant contribution and thus can be considered as a prominent feature in the toolbox of organic chemistry. With respect to aromatic coupling partners, the majority of reports involve indoles; however, other partners do feature in a select few reports. In 2008, Akiyama disclosed an alkylation reaction of indoles catalyzed by phosphoric acid **PA 13.** He showed that nitroalkenes **8a** could couple with N–H-free indoles **8b** to give the corresponding products **8c** with overall excellent selectivities (Scheme 6.12) [19].

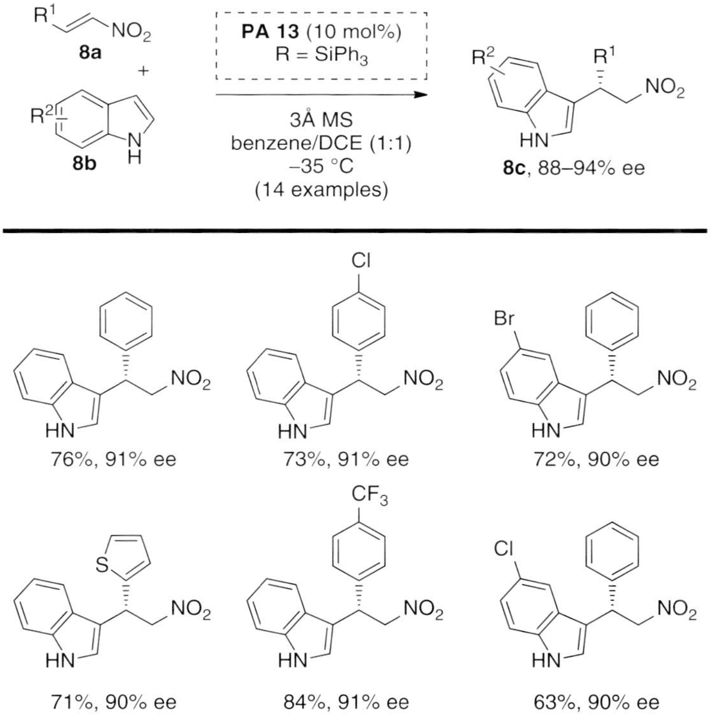

Scheme 6.12 Friedel–Crafts reaction of nitroalkenes by Akiyama [19].

The reaction works well for both aromatic and alkyl substrates, with the latter requiring slightly extended reaction times. In addition, the reaction could easily be scaled up to a gram scale, which delivered an increased yield and an identical level of enantioselectivity of the product.

The proposed mechanism is illustrated in Scheme 6.13. It was shown by the authors that the N–H-free indole is crucial for achieving high yields and selectivity while protected indoles gave suboptimal results. Therefore, it is believed that coordination of the catalyst with both the indole and the nitroalkene occurs during the transition state to yield the desired products. Recently, the process was extended to β,β-disubstituted nitroalkenes to create all carbon quaternary centers [20]. Related studies into this reaction with phosphoric acids have been conducted by Xiao [21] and with bis-sulfonamides by Jørgensen [22].

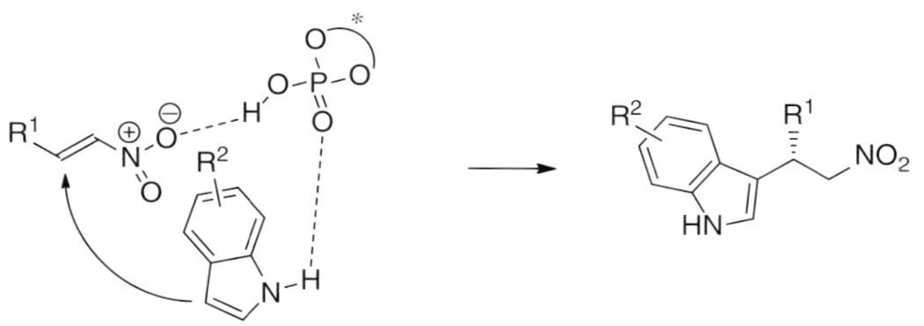

Scheme 6.13 Mechanism of Friedel–Crafts reaction [19].

Also in 2008, Rueping has shown that β,γ-unsaturated α-keto esters **9a** could be activated by *N*-triflylphosphoramide **[H$_8$]-NPA 6** to undergo conjugate addition rather than 1,2-addition by *N*-methyl indoles **9b.** The Friedel–Crafts adducts were obtained with moderate yields and high enantioselectivities (Scheme 6.14) [23].

Worthy of note is that in this protocol, the indole remains protected; however, very good results are still obtained. This is in stark contrast to Akiyama's approach and appears to suggest that mechanistically the indoles may not interact formally with the catalyst. In 2011, Toy used this reaction as a development platform for phosphonium ion tagged catalysts, which would allow a more facile recovery process for possible applications in catalyst recycling. He showed that **PA 35** could be employed in a similar reaction to access the products **9f** with moderate results (Scheme 6.15) [24]. Importantly though, the catalyst could be recovered in greater than 90% yield and when recycled gave slightly lower results. This reaction has also been studied by several other research groups using various chiral phosphoric acid catalysts [25–30]. Notable extensions include the use of α,β-unsaturated acyl phosphonates by Akiyama [31] and an intramolecular variant by You [32].

The use of non-indole components has not received much attention, but some notable pieces of research have been published in this area by the You group. In 2009, You showed that nitroalkenes **10a** can act as partners to 2-substituted pyrroles **10b** to give the adducts **10c** with high enantioselectivities (Scheme 6.16) [33].

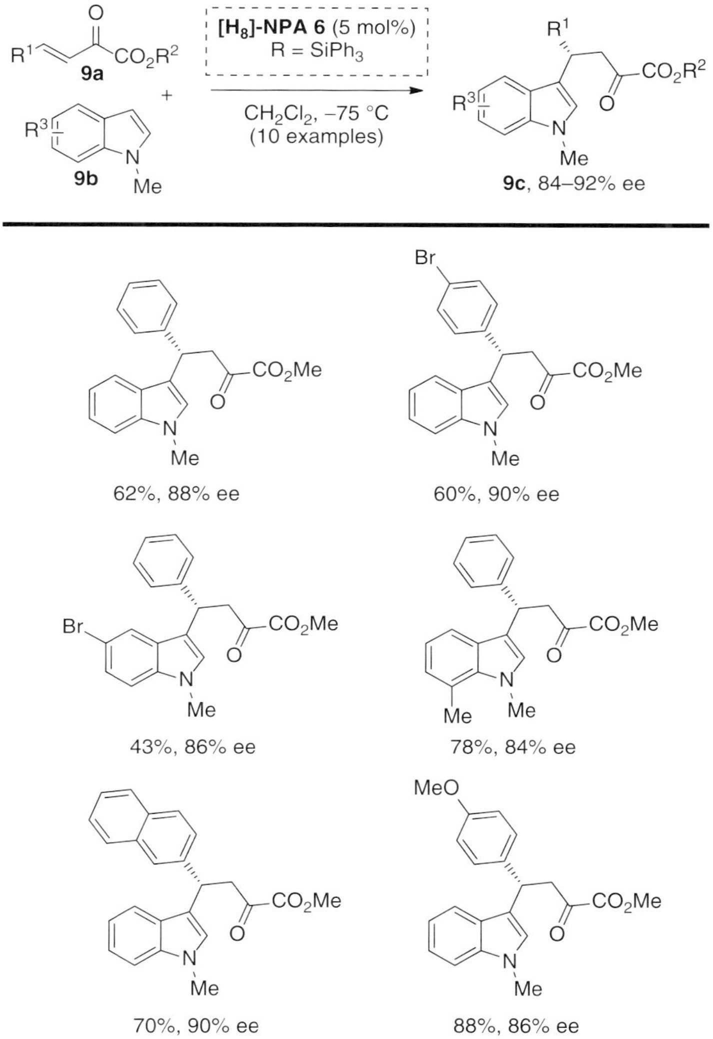

Scheme 6.14 Coupling of β,γ-unsaturated α-keto esters with indoles by Rueping [23].

Following an earlier report, You reported the Friedel–Crafts reaction of 4,7-dihydroindoles, which were shown to undergo reaction at the 2-position rather than the more conventional 3-position seen with indoles. Treating nitroalkenes **10d** with 4,7-dihydoindoles **10e** led to a smooth reaction to give the products **10f** with high enantioselectivities (Scheme 6.17) [34, 35]. The products were shown to be easily oxidized to indoles in a one-pot process, rendering the process complementary to indole reactivity. In addition, the process required an impressively low loading of just 0.5 mol% of **(*S*)-PA 1** to facilitate the reaction.

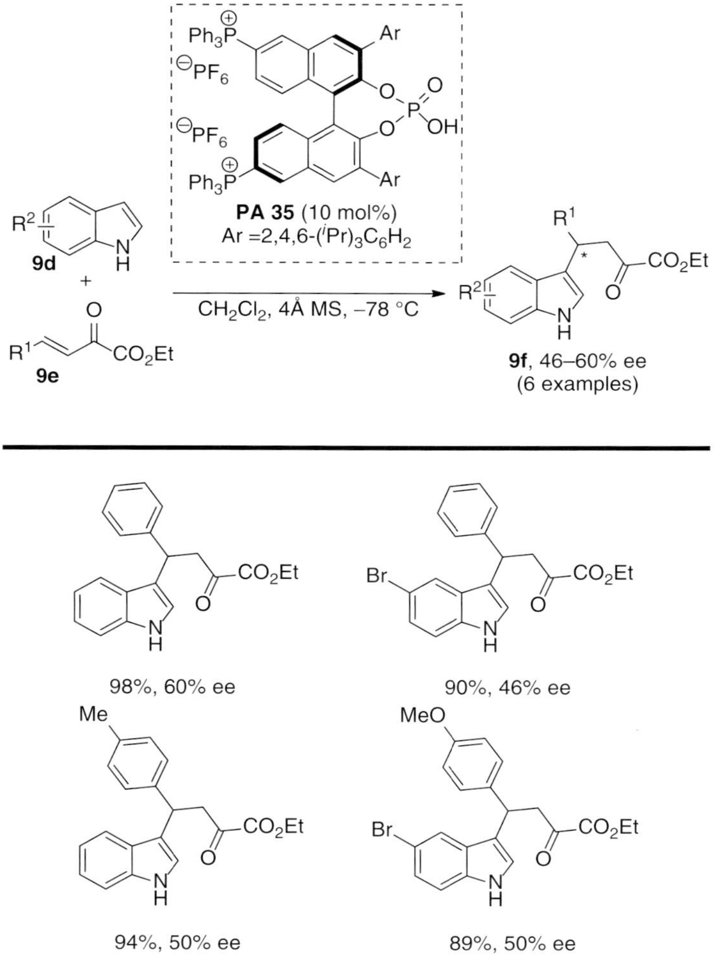

Scheme 6.15 Coupling of β,γ-unsaturated α-keto esters with indoles by Toy [24].

6.3
Pericyclic Reactions

Pericyclic reactions are notoriously difficult to control especially in an asymmetric manner. For Diels–Alder reactions, usually metal Lewis acidic catalysts are employed, which due to their ability to accommodate multiple coordination sites can facilitate such a process to occur. Brønsted acids have shown that in limited reports, they are also able to catalyze such processes with impressive results. A very early report was disclosed by Yamamoto in 2006, who showed that α,β-unsaturated ketone **11a** could be reacted with silyloxydiene **11b** in the presence

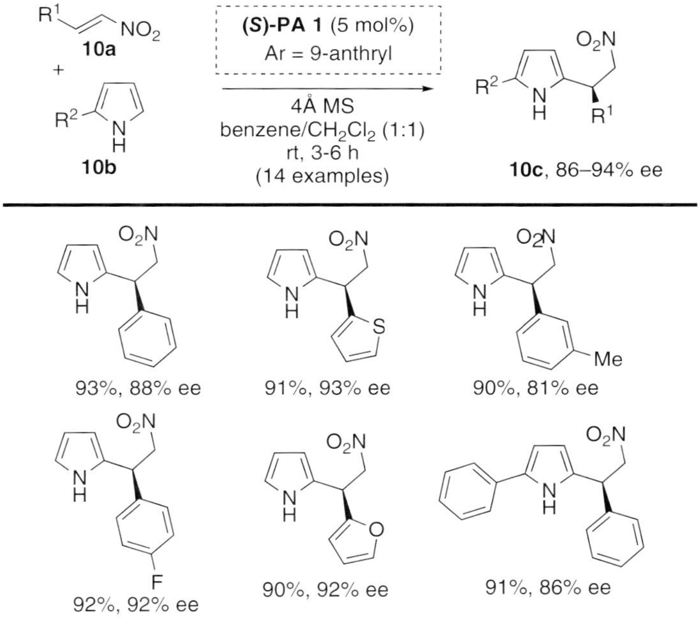

Scheme 6.16 Coupling of nitroalkenes with pyrroles by You [33].

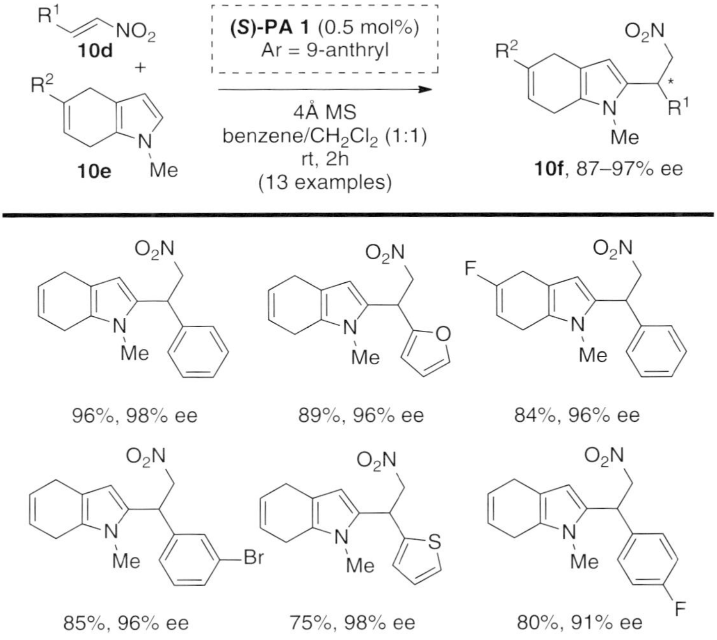

Scheme 6.17 Coupling of nitroalkenes with 4,7-dihydroindoles by You [34].

of N-triflyl phosphoramide **(*S*)-NPA 4** to give the substituted cyclohexanes **11c** with good to high enantioselectivities (Scheme 6.18) [36].

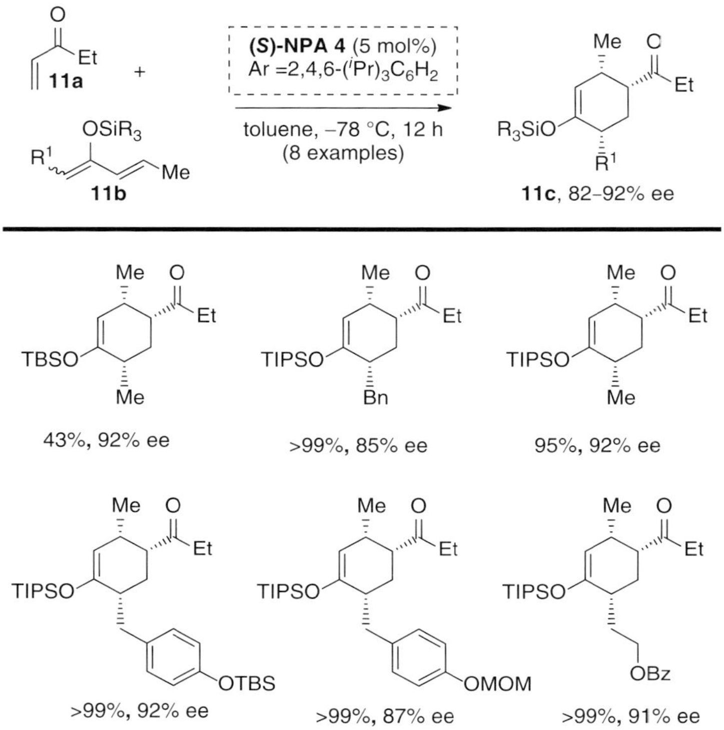

Scheme 6.18 Diels–Alder reaction by Yamamoto [36].

This report holds a significant milestone as it was the first report on the use of an N-triflyl phosphoramide catalyst. The reason for development was crucial since phosphoric acid catalysts were found not to be sufficiently acidic enough for activation of the substrates. In 2013, Nagorny used the same catalyst to perform a similar transformation with α,β-unsaturated acetals [37]. The Diels–Alder reaction also served as a test bed for Terada, who developed a bisphosphoric acid catalyst **PA 36,** which could suitably provide cyclohexanes **11f** with excellent levels of control (Scheme 6.19) [38]. The catalyst, which contains three separate axially chiral axes, was shown to possess internal hydrogen bonding between the two phosphoric acid units, which was proposed to be crucial to achieving activation and selectivity during the reaction. Recently, Melchiorre has disclosed the vinylogous Diels–Alder reaction by activation of cyclic 2,4-dienones [39].

The proposed mechanism of the Diels–Alder reaction is thought to follow standard procedures and can be visualized in Scheme 6.20. If the process is

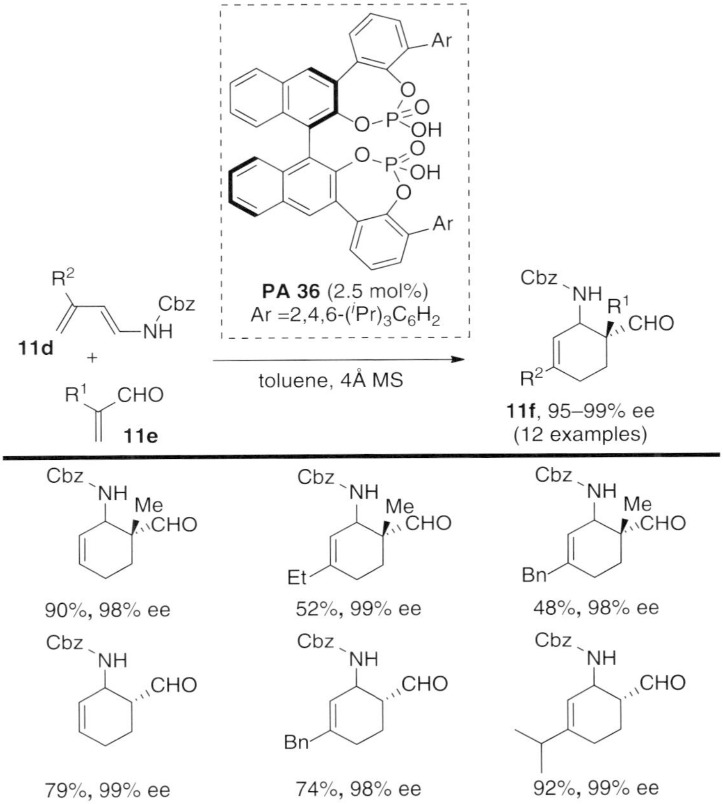

Scheme 6.19 Diels–Alder reaction by Terada [38].

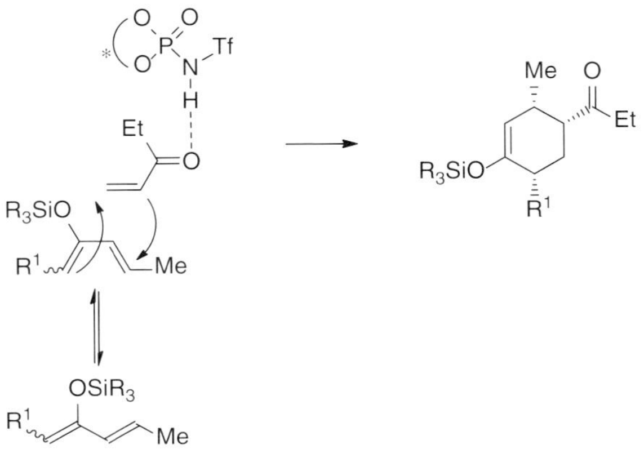

Scheme 6.20 Mechanism of Diels–Alder reaction [36].

concerted, then isomerization of the diene must occur to the s-*cis* form before reacting with the dienophile to yield the desired products. The catalyst is thought to activate the dienophile by hydrogen bonding to the adjacent carbonyl group.

Recently, Zhang has reported on a highly selective [3+2]-coupling of quinone-derived substrates with indoles. He has demonstrated that the reaction between **12a** and **12b** can be facilitated by 5 mol% of **PA 16** to yield the benzofuroindoline **12c** with excellent levels of selectivity (Scheme 6.21) [40].

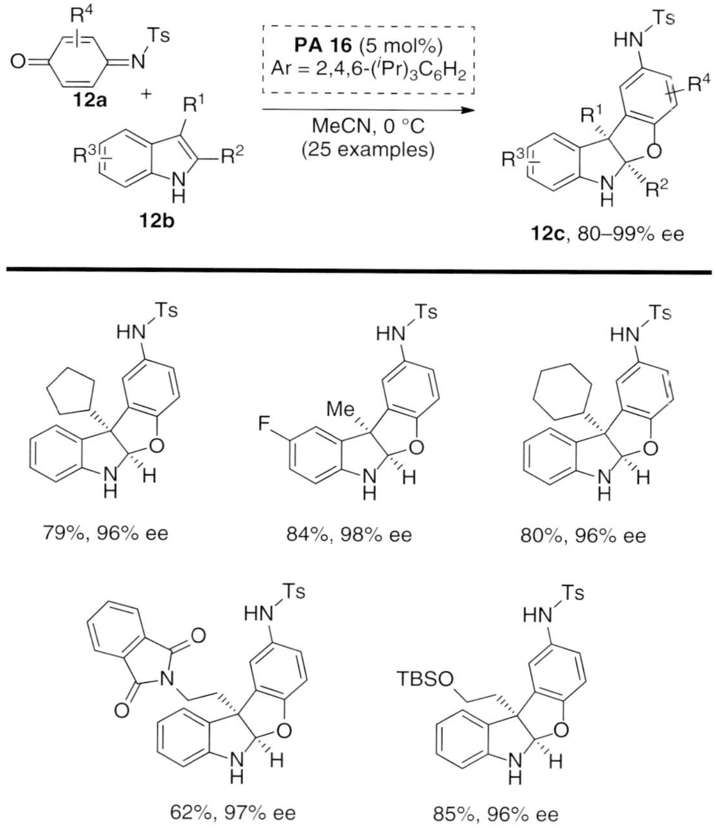

Scheme 6.21 [3+2] coupling of indoles with quinone monoimines by Zhang [40].

Although formally the transformation can be considered a [3+2]-reaction, it is more likely to involve a stepwise process. The proposed mechanism is shown in Scheme 6.22. The authors suggest that the catalyst is involved in activating the quinone monoimine and the indole by hydrogen bonding. This then is suitably reactive enough for attack by the indole at the 3-position followed by concomitant cyclization by the oxygen atom at the 2-position to complete the process. A closely related tandem process using quinone imine ketals has also been disclosed by Shi [41].

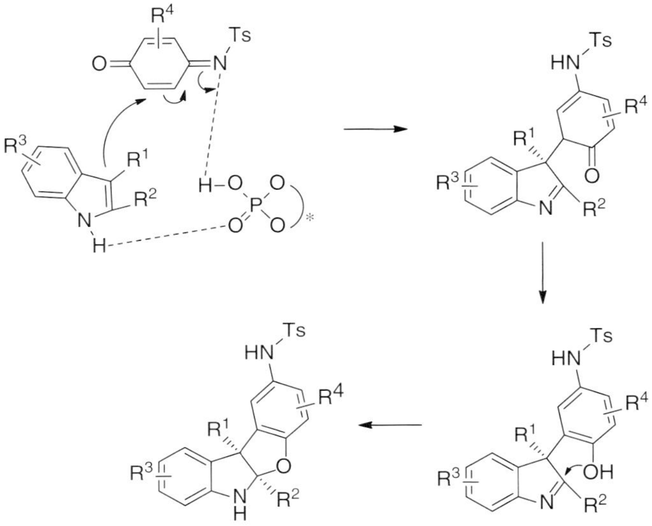

Scheme 6.22 Mechanism of [3+2] coupling of indoles [40].

6.4 Cascades

Cascade reactions are the ideal solution to achieving rapid molecular assembly from simple starting materials, and recent synthetic trends have made this area a highly prominent feature to now being considered a highly prominent feature. On the other hand, Brønsted acids have been shown to facilitate a number of highly useful processes, and so, the unification of Brønsted acids to mediate cascade process is very much an achievable vision. In 2012, Antilla showcased the power of cascade process by using one in the total synthesis of (−)-debromoflustramine B. Firstly, the group developed a cascade process that involved the reaction of various tryptamines with α,β-unsaturated ketones to deliver pyrroloindoline cores, which are also embedded in the target natural product. Subsequently, they then combined α,β-unsaturated ketone **13a** with tryptamine **13b** in the presence of **PA 16** to initiate a directly relevant cascade process to yield the pyrroloindoline **13c** with 91% yield and 93% enantiomeric excess (Scheme 6.23) [42].

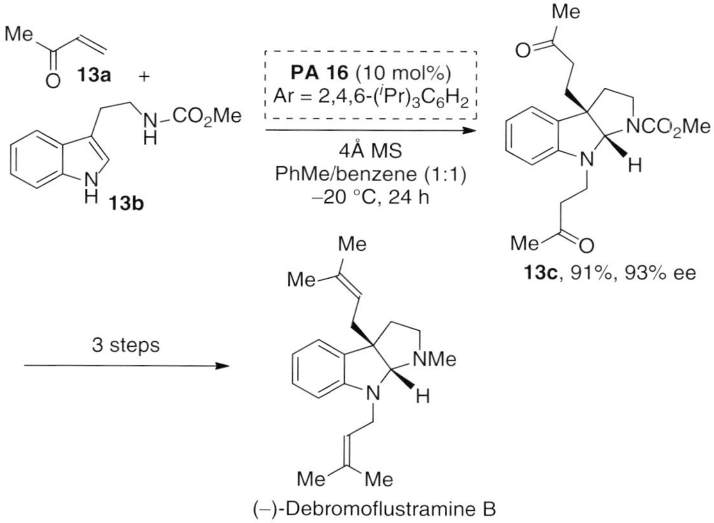

Scheme 6.23 Cascade process in the synthesis of (−)-debromoflustramine B by Antilla [42].

This intermediate then was used to complete the synthesis of (−)-debromoflustramine B by further three straightforward synthetic steps. Additionally, it was shown that α,β-unsaturated ketones could be replaced with diazenes to also yield highly useful products.

The proposed mechanism of the cascade is depicted in Scheme 6.24. The authors suggest that the catalyst activates both the indole proton and the carbonyl of the pendant carbamate. This is in contrast with the previous studies, which would suggest activation of the α,β-unsaturated ketone; however, no NMR shifts were detected when the two components were mixed together. Instead, the carbonyl can be activated by the amide and thus undergo attack by the indole at the C3-position. This generates an iminium ion, which can be subsequently captured by the indole. Finally, addition of another equivalent of the α,β-unsaturated ketone occurs onto the indole to yield the desired products. In the same year, You also developed a comparably similar transformation to access pyrroloindoline derivatives [43].

Polyene cyclizations are considered to be biomimetic cascade reactions that resemble to some extent the powerful transformations that nature is able to perform. The use of acidic conditions to promote these transformations is well known and so the avenue for employing chiral acids is open for exploration. In 2011, Ishihara developed a chiral Lewis-base-assisted Brønsted acid **BA 21** for the development of this methodology (Scheme 6.25) [44].

Having initially explored the scope of the polyene cyclizations, the group pursued the use of this toward the synthesis of bioactive natural products. They showed that by taking **14a** with 40 mol% of **BA 21**, a rather modest yield of 36%

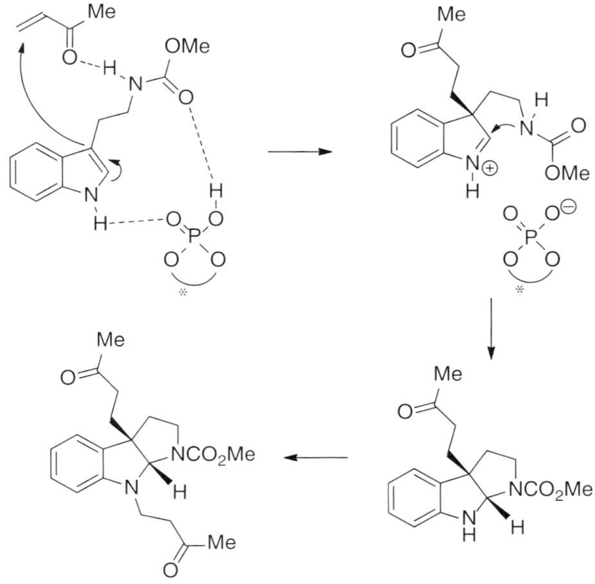

Scheme 6.24 Mechanism of cascade process [42].

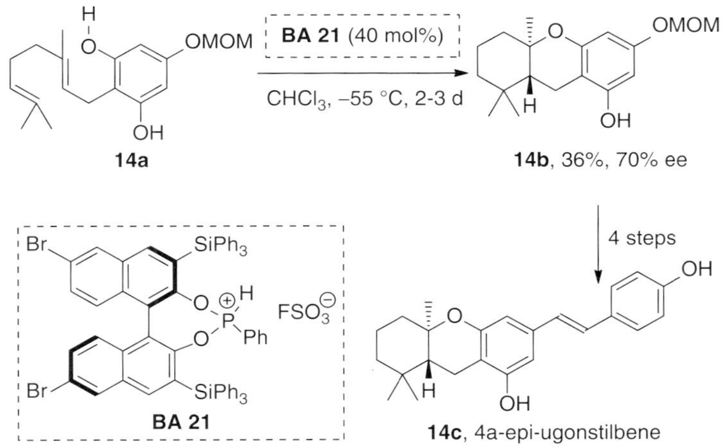

Scheme 6.25 Synthesis of 4a-*epi*-ugonstilbene by Ishihara [44].

was achieved of **14b** but in 70% enantiomeric excess. This could be elaborated to **14c** in four simple steps, which is an epimeric natural product.

References

1. Sugiura, M., Tokudomi, M., and Nakajima, M. (2010) *Chem. Commun.*, **46**, 7799–7800.
2. Sugiura, M., Kinoshita, R., and Nakajima, M. (2014) *Org. Lett.*, **16**, 5172–5175.
3. Ackermann, L. and Althammer, A. (2008) *Synlett*, (7), 995–998.
4. Shapiro, N.D., Rauniyar, V., Hamilton, G.L., Wu, J., and Toste, F.D. (2011) *Nature*, **470**, 245–249.
5. Sun, Z., Winschel, G.A., Zimmerman, P.M., and Nagorny, P. (2014) *Angew. Chem. Int. Ed.*, **53**, 11194–11198.
6. Uraguchi, D., Nakashima, D., and Ooi, T. (2009) *J. Am. Chem. Soc.*, **131**, 7242–7243.
7. Uraguchi, D., Kinoshita, N., Kizu, T., and Ooi, T. (2011) *Synlett*, (9), 1265–1267.
8. Hashimoto, T., Nakatsu, H., Takiguchi, Y., and Maruoka, K. (2013) *J. Am. Chem. Soc.*, **135**, 16010–16013.
9. Zhang, Y.-C., Jiang, F., Wang, S.-L., Shi, F., and Tu, S.-J. (2014) *J. Org. Chem.*, **79**, 6143–6152.
10. Cai, Q., Zheng, C., and You, S.-L. (2010) *Angew. Chem. Int. Ed.*, **49**, 8666–8669.
11. Feng, Z., Xu, Q.-L., Dai, L.-X., and You, S.-L. (2010) *Heterocycles*, **80**, 765–771.
12. Gu, Q., Rong, Z.-Q., Zheng, C., and You, S.-L. (2010) *J. Am. Chem. Soc.*, **132**, 4056–4057.
13. Rueping, M., Moreth, S.A., and Bolte, M. (2012) *Z. Naturforsch., B: Chem. Sci.*, **67**, 1021–1029.
14. Liu, H., Zeng, C., Guo, J., Zhang, M., and Yu, S. (2013) *RSC Adv.*, **3**, 1666–1668.
15. Rueping, M., Uria, U., Lin, M.-Y., and Atodiresei, I. (2011) *J. Am. Chem. Soc.*, **133**, 3732–3735.
16. Zhuang, M. and Du, H. (2014) *Org. Biomol. Chem.*, **12**, 4590–4593.
17. Wang, P.-S., Zhou, X.-L., and Gong, L.-Z. (2014) *Org. Lett.*, **16**, 976–979.
18. Sakuma, M., Sakakura, A., and Ishihara, K. (2013) *Org. Lett.*, **15**, 2838–2841.
19. Itoh, J., Fuchibe, K., and Akiyama, T. (2008) *Angew. Chem. Int. Ed.*, **47**, 4016–4018.
20. Mori, K., Wakazawa, M., and Akiyama, T. (2014) *Chem. Sci.*, **5**, 1799–1803.
21. Lin, J.-H. and Xiao, J.-C. (2011) *Eur. J. Org. Chem.*, **2011**, 4536–4539.
22. Zhuang, W., Hazell, R.G., and Jørgensen, K.A. (2005) *Org. Biomol. Chem.*, **3**, 2566–2571.
23. Rueping, M., Nachtsheim, B.J., Moreth, S.A., and Bolte, M. (2008) *Angew. Chem. Int. Ed.*, **47**, 593–596.
24. Hermeke, J. and Toy, P.H. (2011) *Tetrahedron*, **67**, 4103–4109.
25. Tang, H.-Y., Lu, A.-D., Zhou, Z.-H., Zhao, G.-F., He, L.-N., and Tang, C.-C. (2008) *Eur. J. Org. Chem.*, **2008**, 1406–1410.
26. Scettri, A., Villano, R., and Acocella, M.R. (2009) *Molecules*, **14**, 3030–3036.
27. Pei, Z.-K., Zheng, Y., Nie, J., and Ma, J.-A. (2010) *Tetrahedron Lett.*, **51**, 4658–4661.
28. Gutierrez, E.G., Moorhead, E.J., Smith, E.H., Lin, V., Ackerman, L.K.G., Knezevic, C.E., Sun, V., Grant, S., and Wenzel, A.G. (2010) *Eur. J. Org. Chem.*, **2010**, 3027–3031.
29. Sakamoto, T., Itoh, J., Mori, K., and Akiyama, T. (2010) *Org. Biomol. Chem.*, **8**, 5448–5454.
30. Bi, B., Lou, Q.-X., Ding, Y.-Y., Chen, S.-W., Zhang, S.-S., Hu, W.-H., and Zhao, J.-L. (2015) *Org. Lett.*, **17**, 540–543.
31. Bachu, P. and Akiyama, T. (2010) *Chem. Commun.*, **46**, 4112–4114.
32. Zhang, J.-W., Cai, Q., Shi, X.-X., Zhang, W., and You, S.-L. (2011) *Synlett*, **2011**, 1239–1242.
33. Sheng, Y.-F., Gu, Q., Zhang, A.-J., and You, S.-L. (2009) *J. Org. Chem.*, **74**, 6899–6901.
34. Zeng, M., Kang, Q., He, Q.-L., and You, S.-L. (2008) *Adv. Synth. Catal.*, **350**, 2169–2173.

35. Sheng, Y.-F., Li, G.-Q., Kang, Q., Zhang, A.-J., and You, S.-L. (2009) *Chem. Eur. J.*, **15**, 3351–3354.
36. Nakashima, D. and Yamamoto, H. (2006) *J. Am. Chem. Soc.*, **128**, 9626–9627.
37. Borovika, A. and Nagorny, P. (2013) *Tetrahedron*, **69**, 5719–5725.
38. Momiyama, N., Konno, T., Furiya, Y., Iwamoto, T., and Terada, M. (2011) *J. Am. Chem. Soc.*, **133**, 19294–19297.
39. Tian, X., Hofmann, N., and Melchiorre, P. (2014) *Angew. Chem. Int. Ed.*, **53**, 2997–3000.
40. Liao, L., Shu, C., Zhang, M., Liao, Y., Hu, X., Zhang, Y., Wu, Z., Yuan, W., and Zhang, X. (2014) *Angew. Chem. Int. Ed.*, **53**, 10471–10475.
41. Zhang, Y.-C., Zhao, J.-J., Jiang, F., Sun, S.-B., and Shi, F. (2014) *Angew. Chem. Int. Ed.*, **53**, 13912–13915.
42. Zhang, Z. and Antilla, J.C. (2012) *Angew. Chem. Int. Ed.*, **51**, 11778–11782.
43. Cai, Q., Liu, C., Liang, X.-W., and You, S.-L. (2012) *Org. Lett.*, **14**, 4588–4590.
44. Sakakura, A., Sakuma, M., and Ishihara, K. (2011) *Org. Lett.*, **13**, 3130–3133.

7
Reactions of Other Substrates

So far we have covered many conventionally operating substrates, but this final chapter is dedicated to niche reports of structural motifs, which have not received high levels of attention. Having said that, the reports are no less significant and as chiral Brønsted acids continue to become more versatile, we envision these pioneering works to inspire others to expand this category in the years to come.

7.1
Aziridines

Aziridines are highly useful nitrogen-containing two carbon units and have found plentiful usage in organic synthesis. By virtue of ring strain, aziridines are prone to opening with nucleophiles, and this has been the major focus of methodologies developed with Brønsted acid catalysts. It should be noted, however, that aziridines are not very basic, even less so than regular amines and so discreet strategies to allow their activation by acids are involved in this section and discussed as follows. Antilla has been the dominant user of aziridines, and in 2009, he was the first to show that *meso*-acyl aziridines **1a** could undergo desymmetrization with TMS-N_3 in the presence of **(*S*)-PA 6** (Scheme 7.1) [1].

The transformation leads to the ring-opened products **1b** in good yields and with high enantioselectivities. The group makes use of a VAPOL-derived catalyst **(*S*)-PA 6,** which is shown to provide high levels of selectivity.

Mechanistically, the trimethylsilyl group of TMS−N_3 is shown to be crucial to reactivity (Scheme 7.2). It is proposed that the initial step in this reaction involves the formation of a silylated catalyst, which goes on to activate the acyl group, allowing the stereoselective attack of azide to occur. This observation has also been proposed by Sala, who showed that silylated sulfur nucleophiles could be employed to ring-open *meso*-aziridines [2]. Interestingly, however, in the same year Antilla reported the ring-opening of *meso*-aziridines **2a** with nonsilylated thiols and 10 mol% **(*S*)-PA 6** to give the corresponding ring-opened products **2b** with generally good enantioselectivities (Scheme 7.3, top) [3]. In this case, they propose that a more simple hydrogen-bonding activation of the *N*-acyl motif is responsible for the stereoselectivity observed.

Asymmetric Brønsted Acid Catalysis, First Edition. Magnus Rueping, Dixit Parmar and Erli Sugiono.

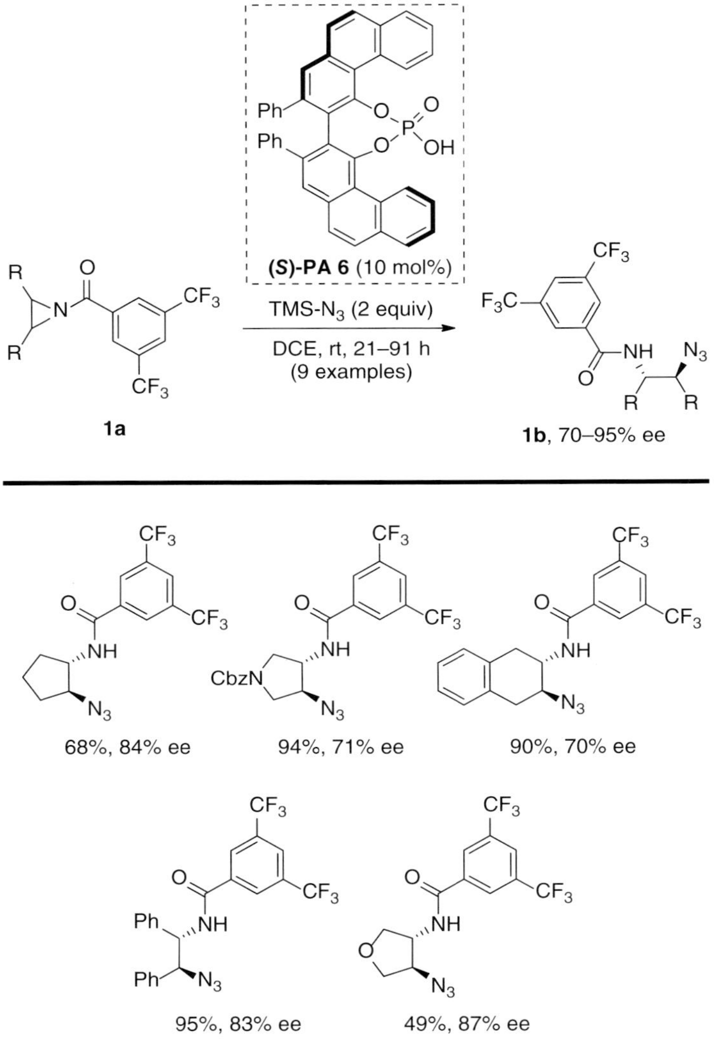

Scheme 7.1 Coupling of *meso*-aziridines with TMS-N_3 by Antilla [1].

In 2011, Della Sala also showed that silylated selenium-derived nucleophiles were suitably for use to ring-open *meso*-aziridines **2a** (Scheme 7.3, bottom) [4]. The group utilized **(*R*)-PA 6,** which delivered the products **2c** generally with high selectivities. Recently, List has also disclosed the opening of aziridines with carboxylic acids; however, it is proposed that the catalyst activates the nucleophile [5].

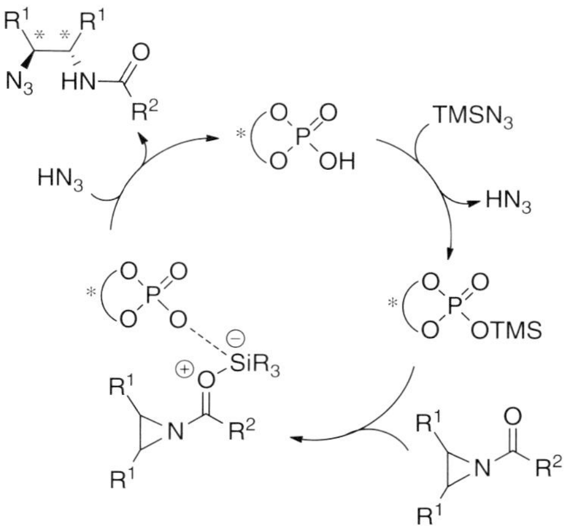

Scheme 7.2 Mechanism of *meso*-aziridines opening [1].

7.2
O-Heterocycles and Ethers

Oxygen heterocycles are notoriously difficult substrates to activate and, in some regards, pose similar challenges that carbonyls provide to Brønsted acid catalysts. The reports in this section are particularly recent and therefore resemble a growing area in this field. Epoxides are substrates that can be considered as comparable to aziridines, which were discussed in the previous section. In 2013, Sun showed that a desymmetrization of epoxides **3a** with aromatic thiols **3b** could be carried out using catalyst **PA 16** in just 2.5 mol% (Scheme 7.4) [6].

The reaction provides the 1,2-difunctionalized products **3c** in excellent yields and with high stereoselectivities. The thiols all possessed a benzothiazole unit, which is proposed to interact with the catalyst through hydrogen bonding in order to achieve the high selectivities observed. Recently, List has also published the opening of epoxides with carboxylic derivatives [7, 8]. Sun has also implemented a similar strategy for the desymmetrizations of 3-substituted oxetanes. He showed that oxetanes **4a** could be opened with benzothiazole containing thiols **4b** to yield the thio-alcohols **4c** with high stereoselectivities (Scheme 7.5) [9].

A very impressive scope was disclosed, and in general, good yields were also obtained of all the products. The products themselves were shown to be useful intermediates for the synthesis of various other building blocks, which also included a formal synthesis of a spermidine alkaloid [10]. The mechanism of the reaction is shown in Scheme 7.6. The catalyst is thought to activate both the thiol and the oxetane moieties in order to facilitate the stereoselective attack and opening of the ring.

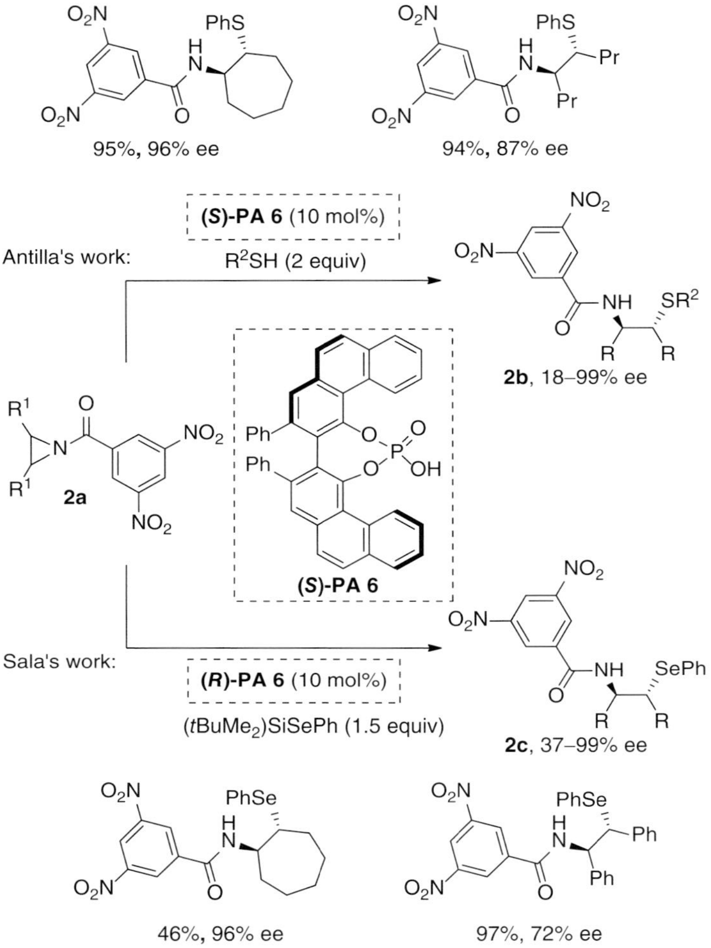

Scheme 7.3 Coupling of thiol- and selenium-derived nucleophiles with *meso*-aziridines by Antilla [3] and Sala [4].

In 2013, epoxides were shown to possess a rather unique behavior toward undergoing an asymmetric rearrangement process. Du demonstrated that aromatic epoxides **5a** could be persuaded to open to yield intermediate **5b,** which underwent an asymmetric hydride shift under the influence of ***N*-SPA 1** and after reduction with $NaBH_4$ gave the products **5c** with modest selectivities (Scheme 7.7) [11].

Catalyst ***N*-SPA 1** was found to be extremely sensitive to the reaction medium, and an optimal balance was found with a 1 : 1 mixture of CH_2Cl_2 and THF between reactivity and stereoselectivity. The subsequent reduction step was shown to be able to be performed in a one-pot procedure without any loss of efficiency.

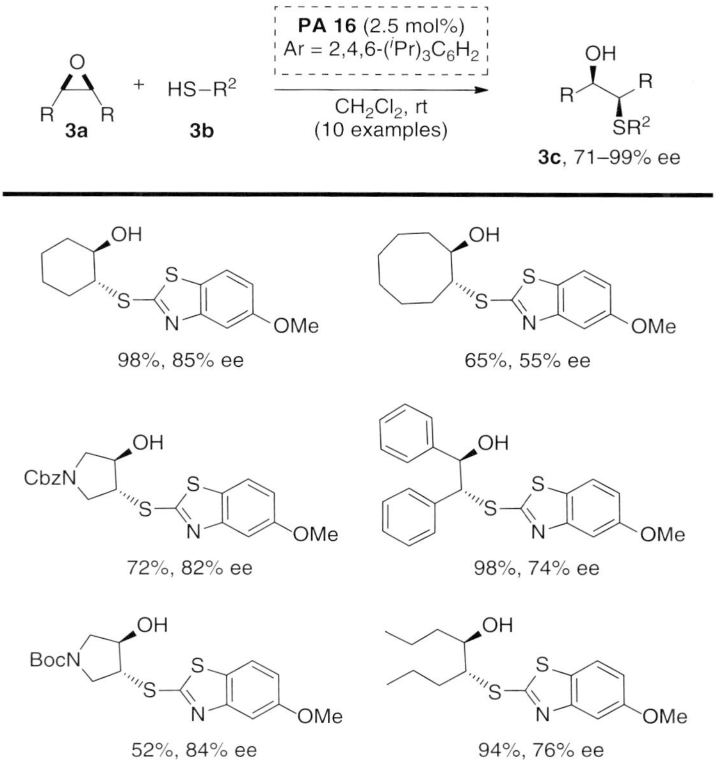

Scheme 7.4 Desymmetrization of *meso*-epoxides by Sun [6].

Activation modes of substrates by Brønsted acid catalysts are constantly evolving and the development of new modes is what has driven this field to the level of prominence it has today. In 2013, the List group developed a new activation mode, which involved the direct attack of nucleophiles at sp^3-hybridized centers to perform kinetic resolutions mediated by a chiral phosphoric acid. More specifically, he showed that benzhydryl ethers ***rac*-6a** would selectively undergo cyclative displacement to yield **6b** with good to high stereoselectivity factors (Scheme 7.8) [12].

The resolution is thought to involve the preferential activation of one enantiomer of **6a** in undergoing a S_N2-type alkylation by the pendant benzylic alcohol group. The direct attack of nucleophiles on sp^3-hybridized centers is scarce and so this development resembles an important milestone. The mechanism can be observed in Scheme 7.9. The catalyst is proposed to activate both the leaving group and the pendant alcohol group toward undergoing direct attack. Worthy of note is that the reaction can be carried out at temperatures as high as 120 °C while still maintaining a high level of selectivity.

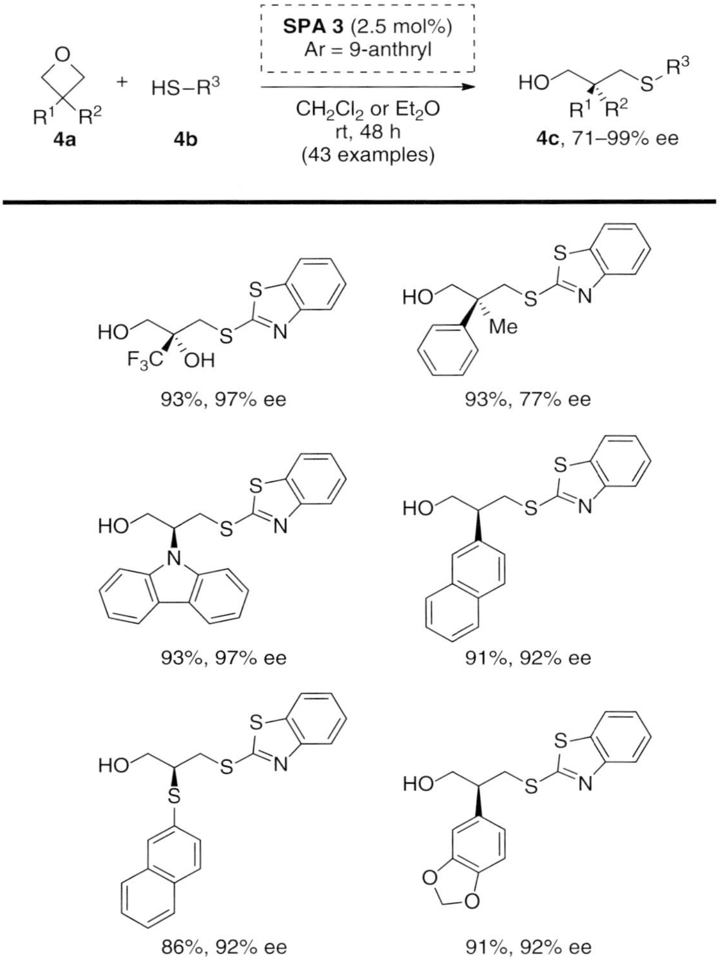

Scheme 7.5 Desymmetrization of oxetanes by Sun [9].

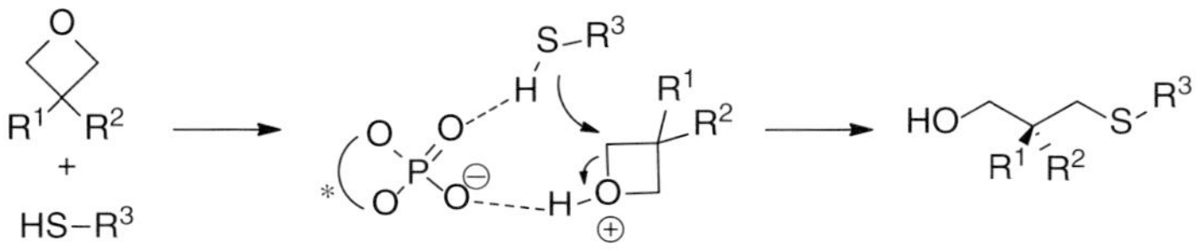

Scheme 7.6 Mechanism of desymmetrization of oxetanes [9].

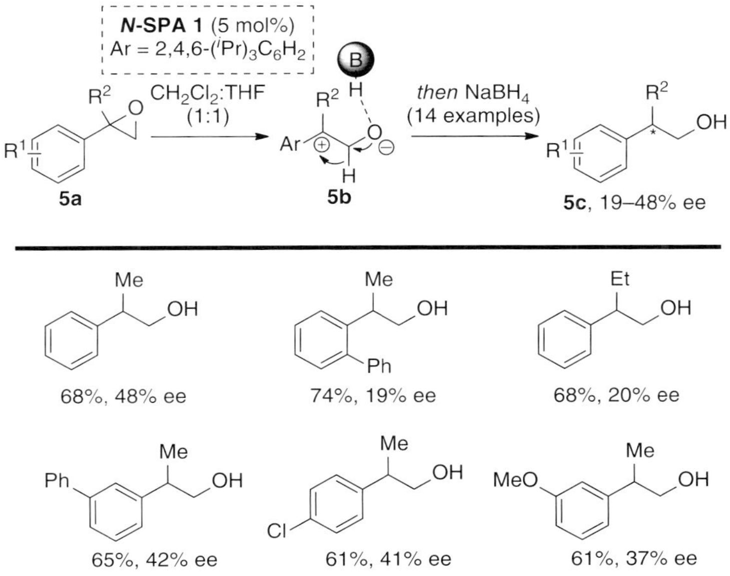

Scheme 7.7 Rearrangement of racemic epoxides by Du [11].

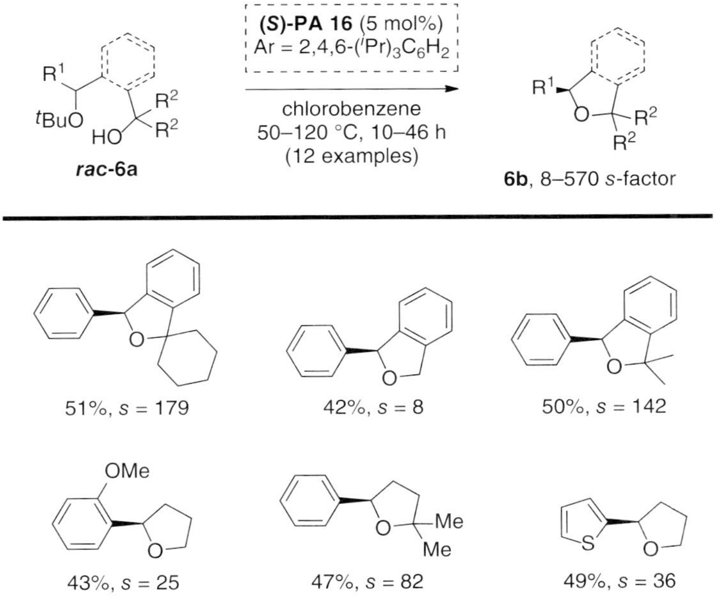

Scheme 7.8 Resolution with S_N2-type O-alkylations by List [12].

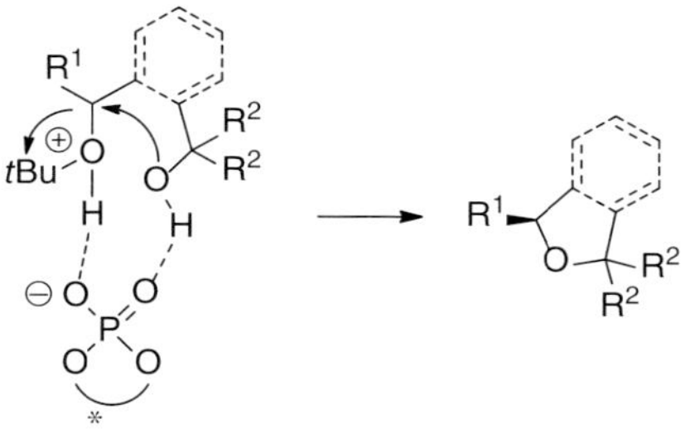

Scheme 7.9 Resolution with S_N2-type O-alkylations by List [12].

7.3
Hydrazines and Hydrazones

Rearrangements are powerful synthetic methodologies, which open up access to chemical motifs that can be difficult to reach using conventional procedures. Although a vast array of rearrangements are known, they are typically difficult to

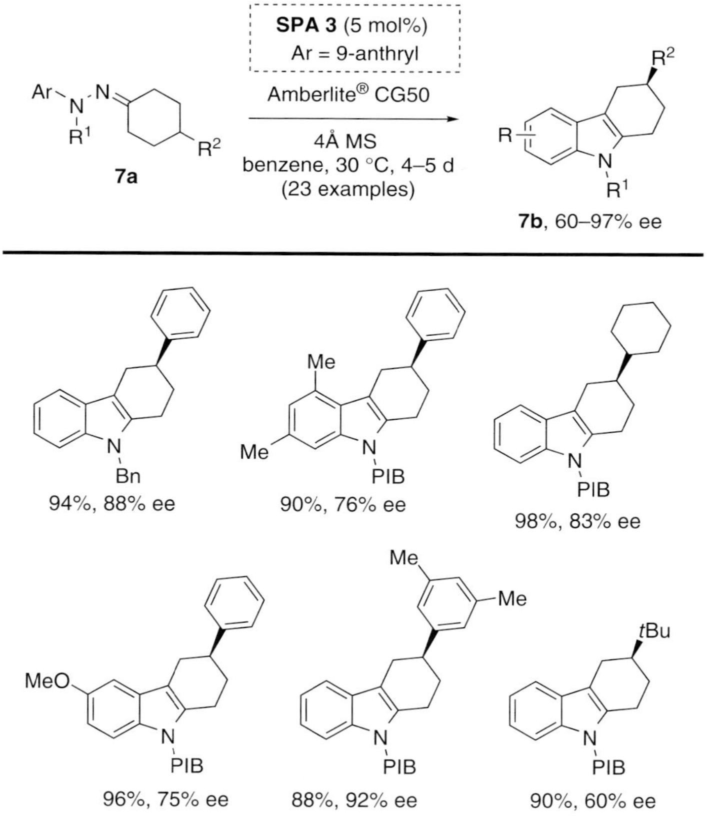

Scheme 7.10 Fischer indole reaction by List [13].

control as they can involve complex transition states and/or high temperatures, which chiral catalysts struggle to influence. The Fischer indole synthesis is a routinely employed procedure for the synthesis of indole derivatives, but the first report of a catalytic asymmetric variant was only provided for the first time by the List group in 2011 using a chiral spirocyclic phosphoric acid catalyst. The group showed that by using aryl hydrazones **7a** with the weakly acidic ion exchange resin Amberlite CG50 and catalyst **SPA 3,** the desired reaction proceeded to give the indole products **7b** with high enantioselectivities (Scheme 7.10) [13].

The resin's role is to scavenge the ammonia by-product produced during the reaction and thus release the catalyst to further perform the desired transformation. The mechanism is depicted in Scheme 7.11. Formally, the process is a desymmetrization, and the group proposes that as the catalyst protonates the hydrazone, one diastereomeric ion pair undergoes an irreversible rearrangement at a higher rate and hence results in the enantioselectivity observed.

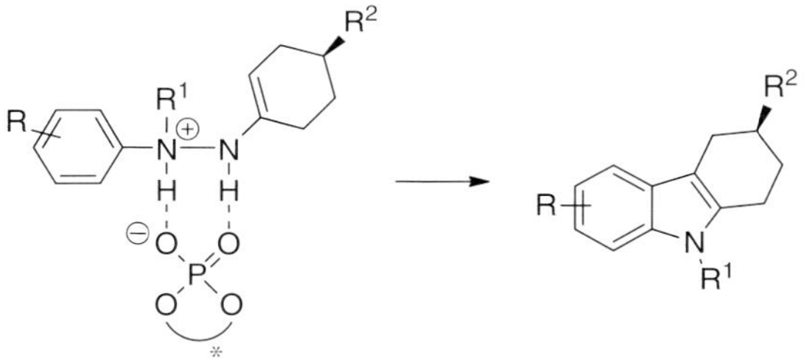

Scheme 7.11 Mechanism of Fischer indole reaction [13].

The List group has also employed this strategy to various related transformations [14, 15]. A closely related [3,3]-rearrangement of hydrazines has also been reported by the groups of Kürti and List independently. Kürti demonstrated that hydrazines **8a** with catalyst **PA 18** could undergo an analogous nitrogen Cope rearrangement to yield 2,2-diamino-1,1′-binaphthalenes **8b** with generally high enantioselectivities (Scheme 7.12) [16].

List's work utilized the same catalyst at 5 mol% in a chlorinated solvent, which transformed **8c** into **8d** at −50 °C with high enantioselectivities (Scheme 7.13) [17]. The weakly acidic resin CG-50 was employed to allow for lower catalyst loadings.

The mechanism of the process is illustrated in Scheme 7.14. Initial mechanistic studies observed a negative nonlinear effect for the rearrangement. The group propose that actually 2 equiv. of the catalyst are involved in activating the starting material by a double protonation. This dicationic species then can undergo a [3,3]-rearrangement and, after proton transfer, yield the desired products.

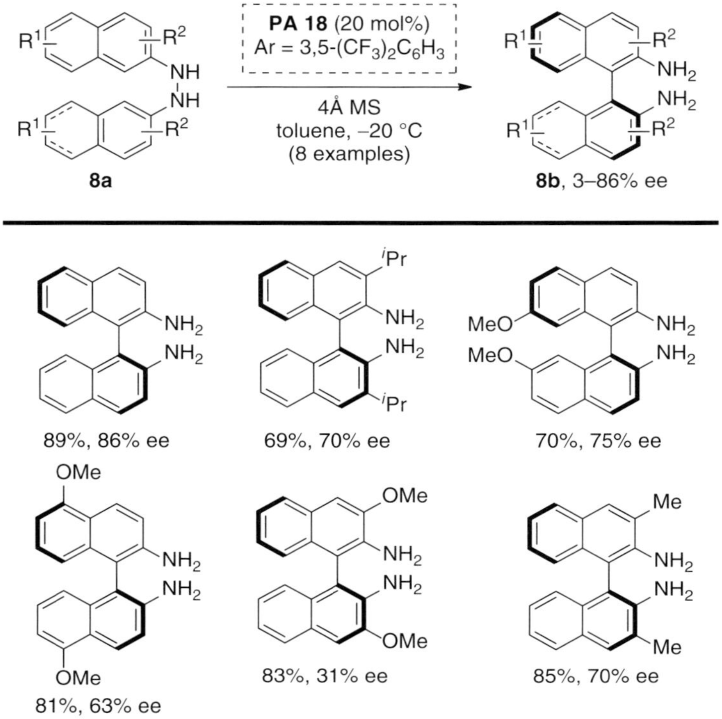

Scheme 7.12 Benzidine rearrangement by Kürti [16].

7.4
Azo/Diazo Substrates

Azo substrates are compounds containing linked nitrogen atoms with substitution on both sides, which usually deems them electrophilic. They are a highly useful source of nitrogen, especially since the *N,N*-linkage can easily be broken by conventional methods. In Section 6.4, we discussed Antilla's use of tryptamine as a nucleophile for alkenes. In the same report, Antilla showed that tryptamines **9a** could also be used as suitable partners for diethyl diazene-1,2-dicarboxylate **9b** in the presence of **PA 16** (Scheme 7.15) [18].

The cascade process leads to the aminated products **9c** in good yields and with high enantioselectivities. The N–N bond was shown to be easily cleaved without any loss of enantiopurity. The mechanism is given in Scheme 7.16. It is thought that the catalyst is coordinated to the pendant amide carbonyl while activation of the diazo starting material occurs by the amide's nitrogen. Subsequent attack followed by concomitant cyclization leads to the products. More recently, Toste

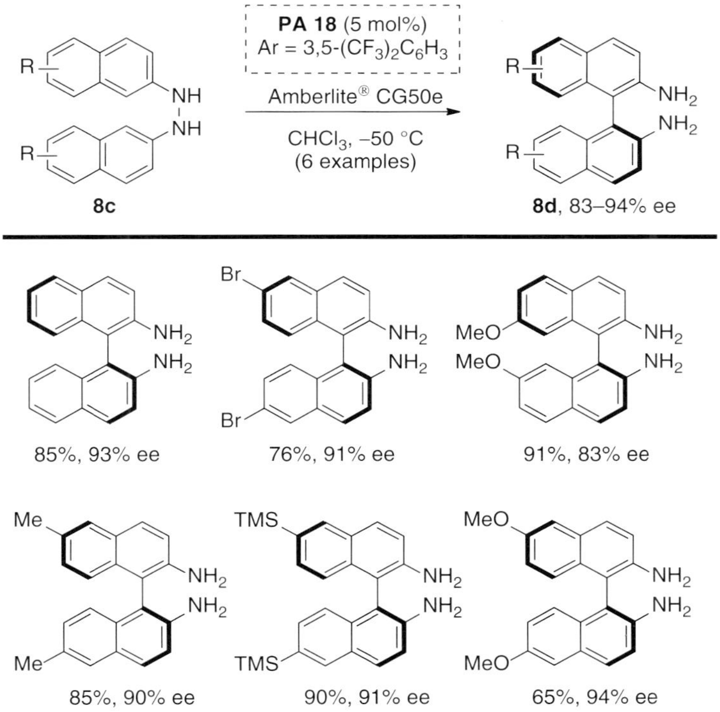

Scheme 7.13 Benzidine rearrangement by and List [17].

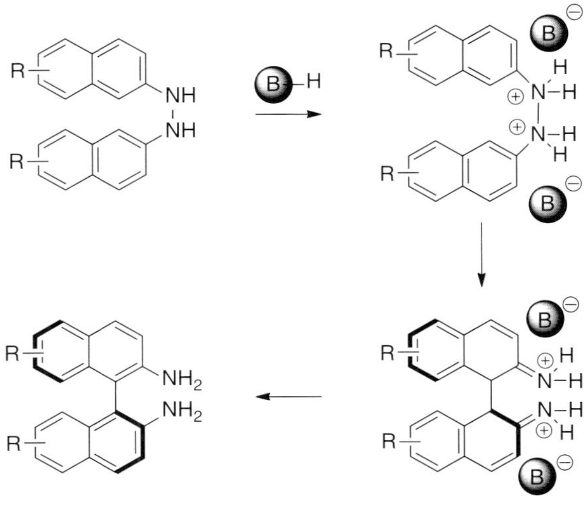

Scheme 7.14 Mechanism of benzidine rearrangement [17].

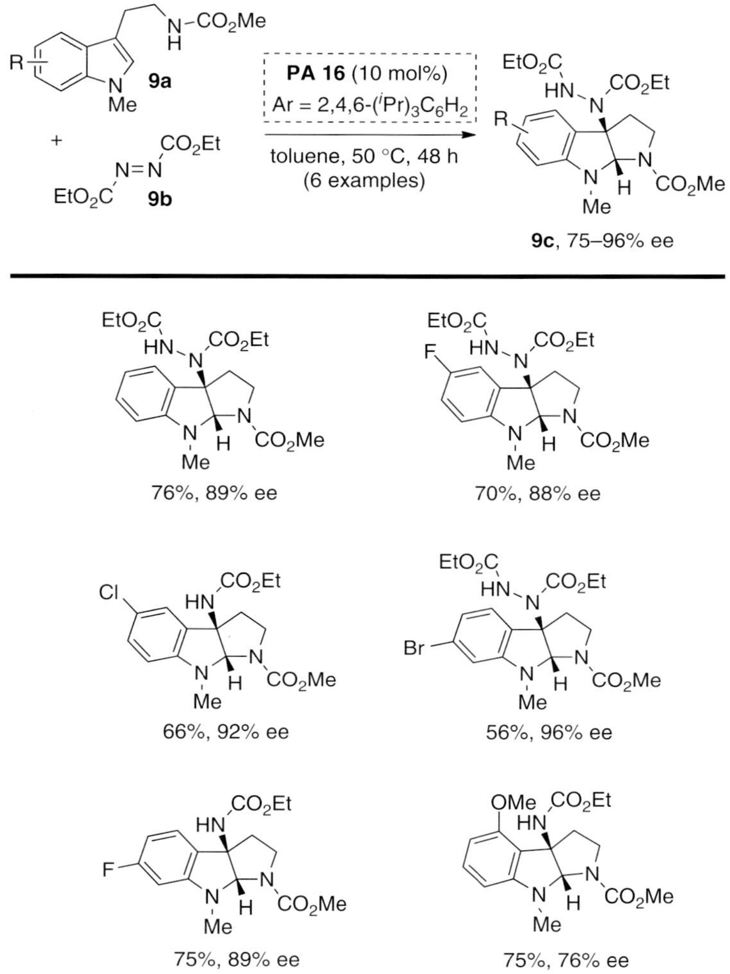

Scheme 7.15 Amination of tryptamines by Antilla [18].

[19, 20] and other research groups [21, 22] have employed azo-substrates with enolate precursors.

Diazo substrates also possess a nitrogen–nitrogen bond but only contain a terminal group, which results in them behaving as both a nucleophile and an electrophile at the terminal position. Typically, these substrates can be activated by metal catalysts, but in 2012, Maruoka found that dicarboxylic acid catalyst **BA 22**

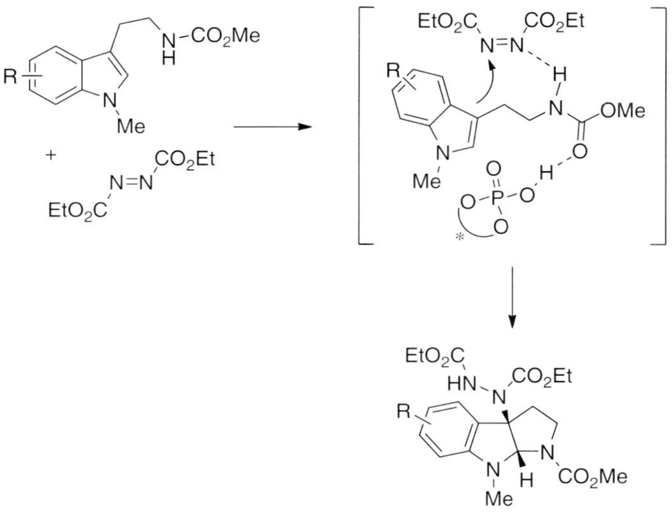

Scheme 7.16 Mechanism of amination [18].

was able to perform desymmetrizing semi-pinacol rearrangements of **10a** using toluene containing an equivalent of H_2O (Scheme 7.17) [23].

To aid the isolation and enantioselectivity determination, a Krapcho decarboxylation using LiCl was employed to remove the ester group and thus provide access to chiral heptanones **10b** with good enantioselectivities. The proposed mechanism of the reaction is given in Scheme 7.18. In this reaction, the exact role of the phosphoric acid is unclear, but it may be involved in activating the diazo-group by providing a proton source as well as secondary attractive interactions that facilitate the rearrangement process.

The use of diazo-esters has also recently been shown by Mattson. She found that nitro diazo-esters **11a** were susceptible to the addition of two nucleophilic components, namely amines **11b** and indoles **11c** in the presence of 20 mol% of **PA 4** (Scheme 7.19) [24].

The reaction proceeds to deliver the products **11d** with modest levels of selectivity but as with the previous report provides a breakthrough in metal-free catalysis of diazo-compound reactions. The mechanism is depicted in Scheme 7.20. The mechanism is, however, unclear, but it is thought that the initial reaction is N–H insertion into the diazo-compound compound and loss of N_2. It is likely that this step is stereoselective and the subsequent displacement of the nitro group occurs with inversion of stereochemistry. However, interestingly, when an *N*-protected indole is used, the enantioselectivity is lost, suggesting hydrogen-bonding interactions between itself and the catalyst.

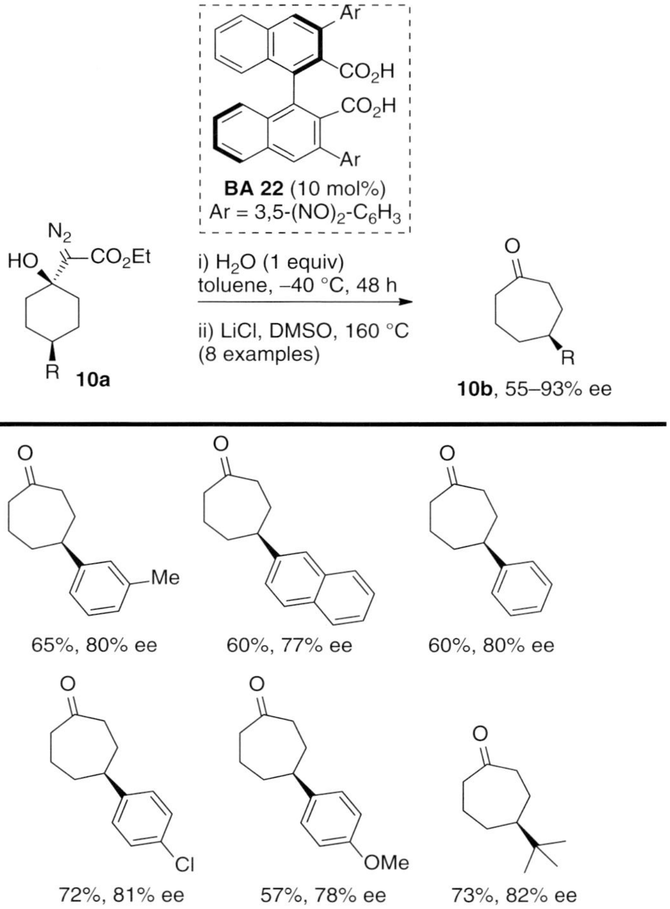

Scheme 7.17 Semi-pinacol rearrangements by Maruoka [23].

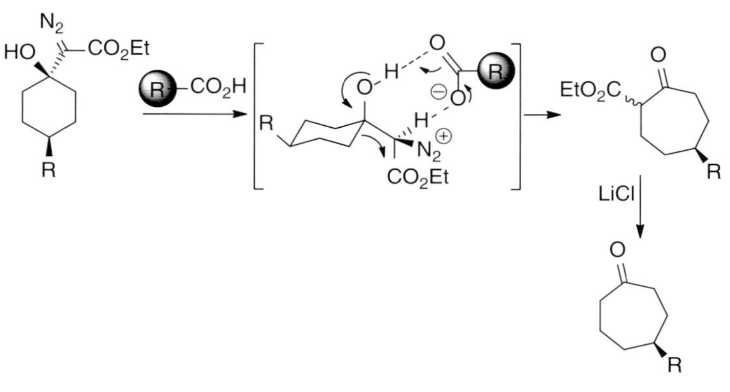

Scheme 7.18 Mechanism of semi-pinacol rearrangements [23].

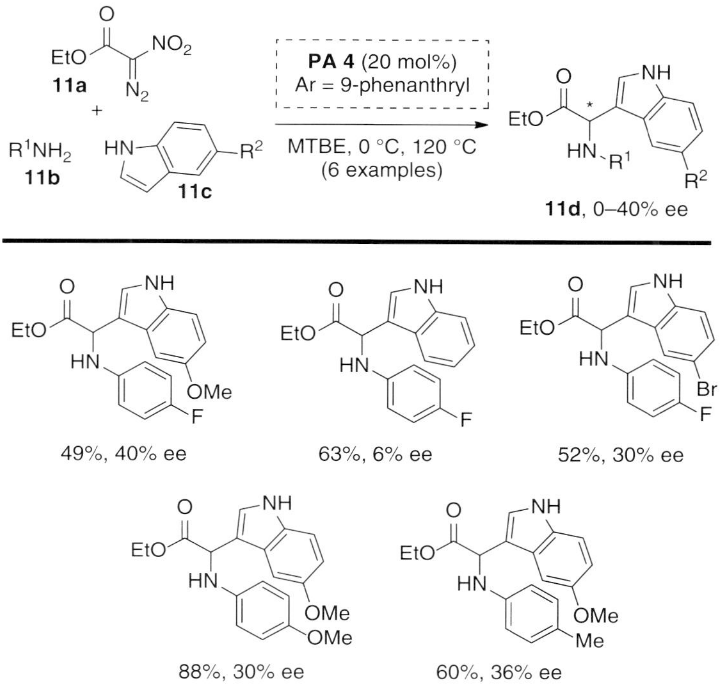

Scheme 7.19 Double addition to nitrodiazoesters by Mattson [24] .

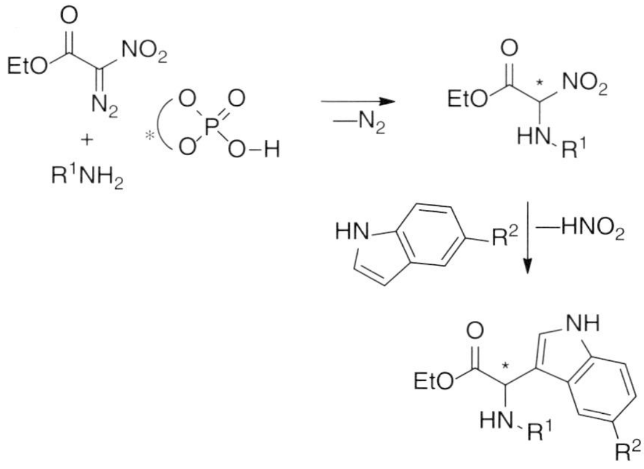

Scheme 7.20 Mechanism of double addition to nitrodiazoesters [24].

7.5
Halogens

7.5.1
Fluorine

The incorporation of fluorine has attracted the attention of many research groups as its importance especially in pharmaceutical and agrochemicals has been recognized over the past few decades. Controlling fluorination reactions especially asymmetric variants is a particularly challenging task as most fluorine sources are highly reactive. This area received a breakthrough from the Toste group in 2011, who demonstrated that a highly lipophilic phosphoric acid catalyst could be employed as a chiral anionic phase-transfer agent for the cationic electrophilic fluorinating agent Selectfluor. The catalyst developed was **PA 37,** which contained

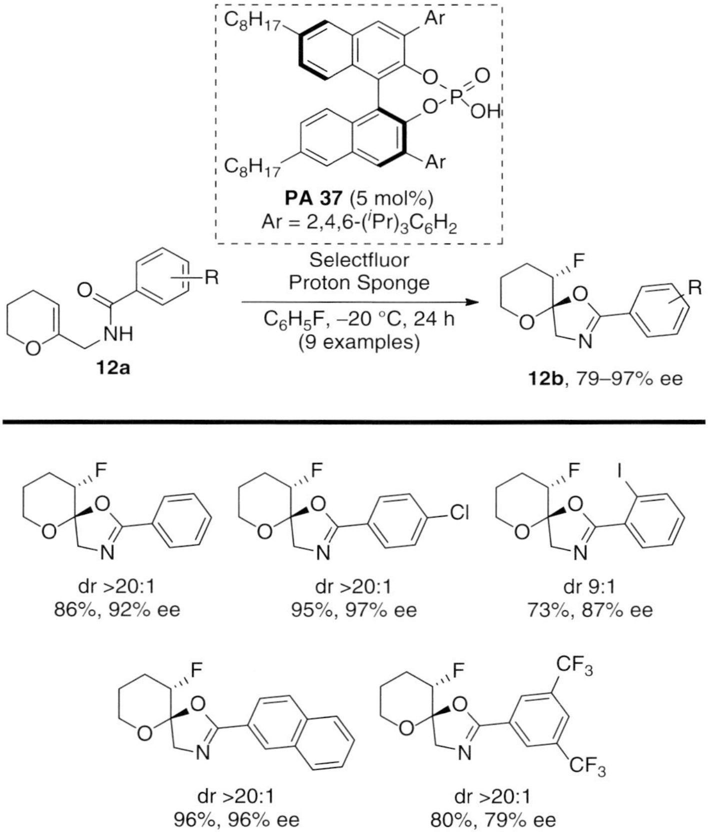

Scheme 7.21 Electrophilic fluorination using phase-transfer catalyst by Toste [25].

sterically hindered groups at the 3,3′-positions and long chain alkyl groups at 7,7′-positions. This was shown to function as a chiral anion toward Selectfluor to perform the electrophilic fluorination and subsequent cyclizations of enol ethers **12a** to yield oxazolines **12b** (Scheme 7.21) [25].

The reaction proceeds smoothly for a range of enol ethers and could also be extended to aromatic alkenes with great effect. Judicial choice of the reaction medium was crucial in order to achieve the desired phase-transfer effect, with fluorobenzene being found to be the optimal choice. A Proton Sponge was included in the reaction to regenerate the catalyst as well as neutralizing any acid produced during the reaction.

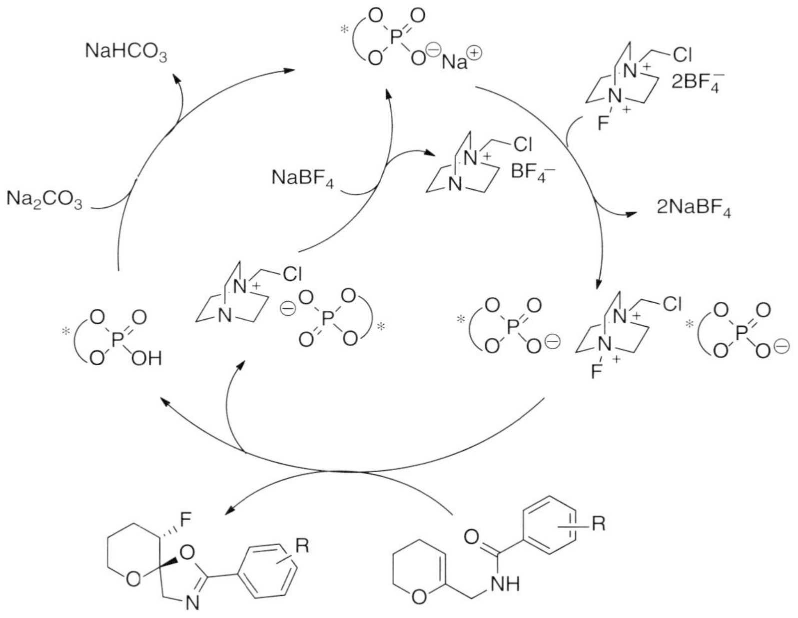

Scheme 7.22 Mechanism of phase-transfer catalysis [25].

The mechanism of this exquisite process is illustrated in Scheme 7.22. It is thought that the active catalytic species is an ionic complex involving 2 equiv. of the catalyst with Selectfluor. Evidence of this was obtained by the observation of nonlinear effects. This complex can then go on to stereoselectively deliver F^+ to the enol ether, which is followed by concomitant cyclization of the amide. The catalyst is then regenerated by deprotonation by the inorganic base. Subsequently, phenols have been shown to be suitable nucleophiles [26] and transformations involving bromo- and iodo-Selectfluor type reagents have also been developed for a variety of reactions [27–29].

This system has proved to be incredibly versatile, and this method of enantioselective incorporation of fluorine has been shown to function well with a variety of different substrates. In 2012, Toste showed that it could be used for the direct fluorination of enamines. By treating a series of cyclic and acyclic enamines **13a** with Selectfluor and catalyst **(*S*)-PA 37,** a smooth fluorination takes place to yield α-fluoroimines **13b** with high enantioselectivities (Scheme 7.23) [30].

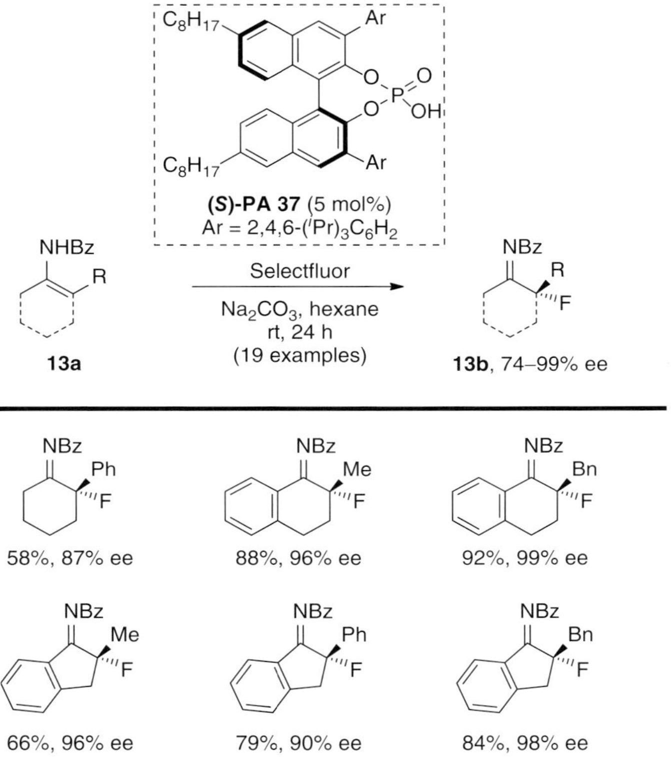

Scheme 7.23 Fluorination of enamides by Toste [30].

Although the products might be expected to be unstable, by the use of a benzoyl protecting group, the ketimine products were stable enough to be easily isolated. The resulting products themselves are highly versatile intermediates and could be amenable to further transformations. It was later shown that further transformations could also be performed in the same reaction pot [31]. An interesting application of this phase-transfer system was shown in 2013, which involved a 1,4-aminofluorination reaction of conjugated dienes **14a** (Scheme 7.24) [32].

The phase-transfer conditions allow for the reaction to proceed as homogeneous conditions showed incompatibilities between Selectfluor and the substrates. The reaction proceeds in a highly regioselective manner to deliver the fluoroamination products **14b** with very high enantioselectivities. The process is thought to be occurring in a concerted fashion with the nitrogen nucleophile

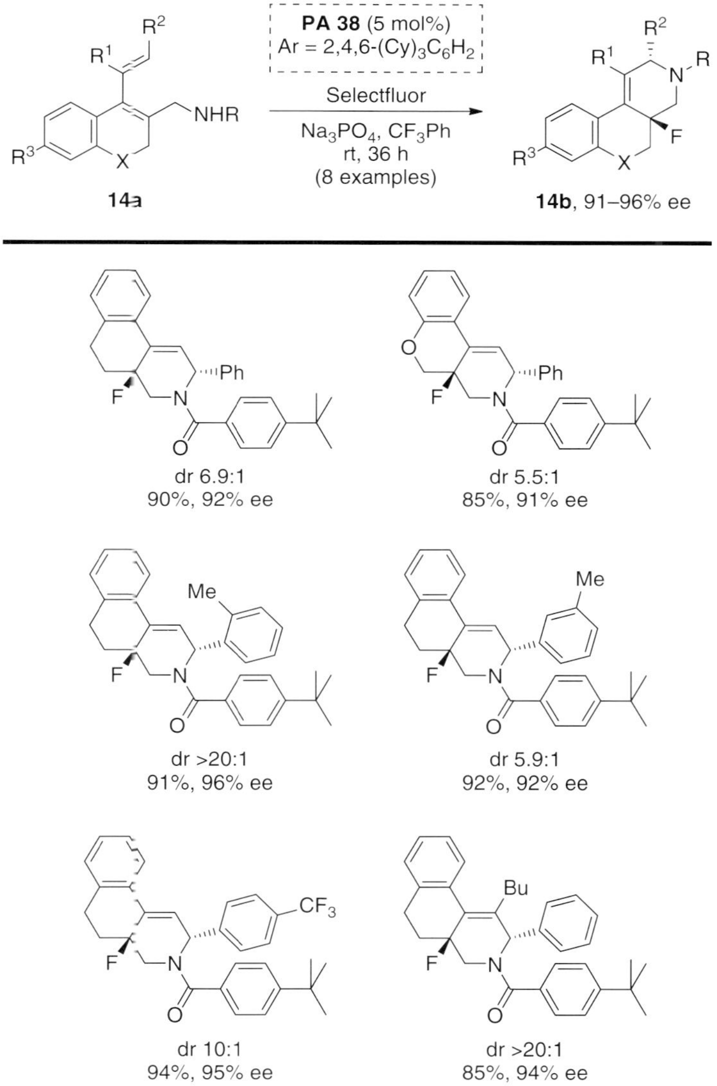

Scheme 7.24 Fluoroamination using phase-transfer catalyst by Toste [32].

also acting as a directing group for the fluorination. The concept of directed fluorination has also been studied in other related systems [33, 34].

The potential of phase-transfer fluorinations with chiral lipophilic phosphoric acids has also attracted the attention of other research groups. The Alexakis group has recently disclosed a Wagner–Meerwein rearrangement of rather strained allylic alcohols **15a** by using Selectfluor and catalyst **PA 39** (Scheme 7.25) [35].

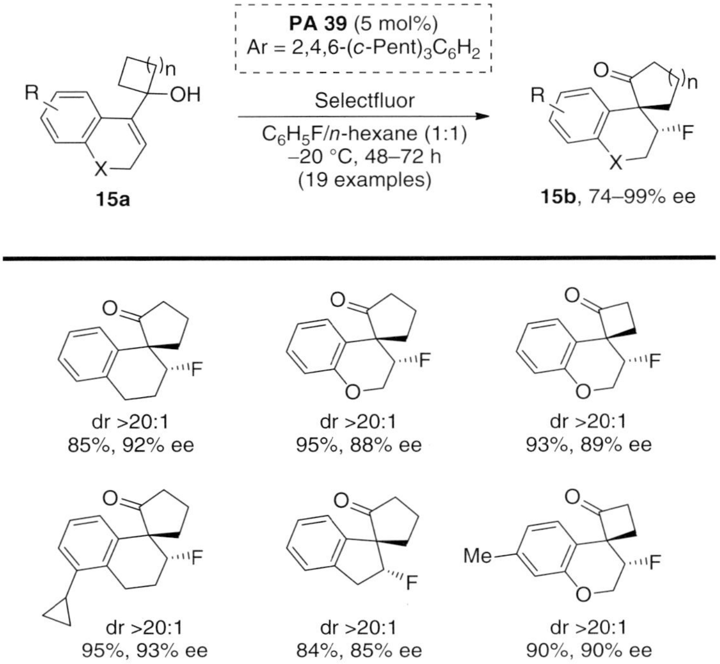

Scheme 7.25 Fluorination-induced Wagner–Meerwein rearrangement by Alexakis [35].

As seen previously, the choice of solvent was crucial in order to facilitate the phase-transfer process and achieve high levels of enantioselectivity for the products **15b.** A 1 : 1 mixture of fluorobenzene and hexane worked best for the highly lipophilic catalyst **PA 39**.

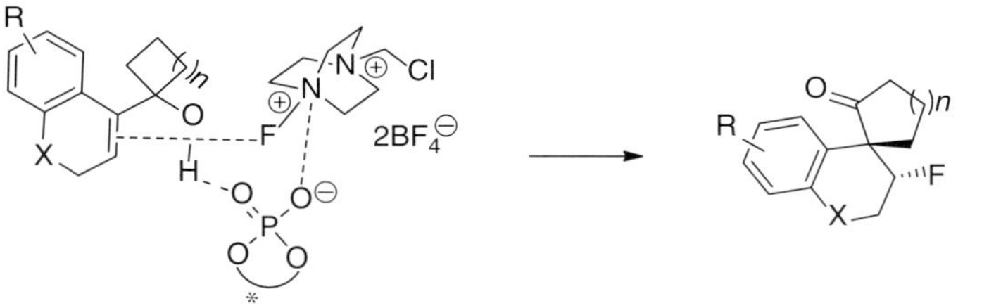

Scheme 7.26 Mechanism of Wagner–Meerwein rearrangement [35].

The mechanism of the transformation is shown in Scheme 7.26. It is thought that a phosphate-Selectfluor complex is responsible for the enantioselectivity by hydrogen-bonding interactions to the alcohol group, which initiates the desired rearrangement. The Alexakis group has also studied similar systems for performing a variety of related transformations [36–38].

7.5.2 Bromine

The introduction of bromine (and iodine) into molecules can be a highly useful process since they are versatile leaving groups for stereoselective displacement by nucleophiles. In the previous subsection, we referred to bromine and iodine Selectfluor-like reagents, and so, these are not discussed again in this section as they are employed in a phase-transfer manner similar to Selectfluor. In fact, the reactivity of bromine sources is usually controlled and sufficiently reactive enough. However, the combination of these sources with chiral Brønsted acids has only received a modest level of attention. Akiyama disclosed an interesting use of a chiral phosphoric acid catalyst to catalyze the kinetic resolution of biaryls through an asymmetric bromination reaction. He showed that achiral biaryls **16a** could be firstly brominated by *N*-bromophthalimide (NBP) in a desymmetrization reaction by catalyst **[H$_8$]-PA 1,** and then via a kinetic resolution, the minor isomer was removed by a second bromination leaving behind **16b** with good to high enantioselectivites (Scheme 7.27) [39–41].

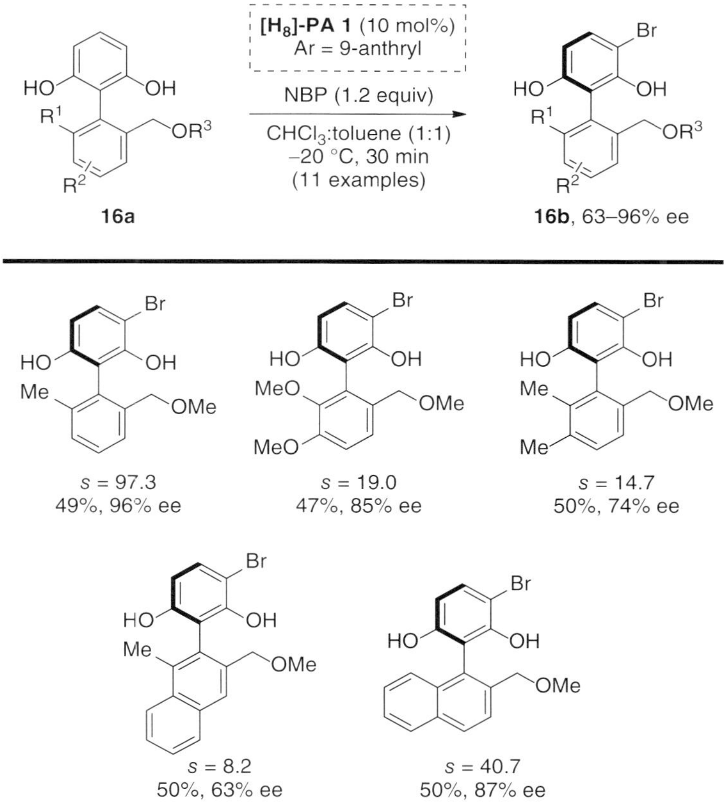

Scheme 7.27 Kinetic resolution of bi-aryls by Akiyama [39].

The mechanism of the process is illustrated in Scheme 7.28. The group performed a series of computational and mechanistic experiments, which suggested that a series of highly ordered hydrogen bonding connections between the substrate, catalyst, and brominating agent is responsible for the selectivity observed.

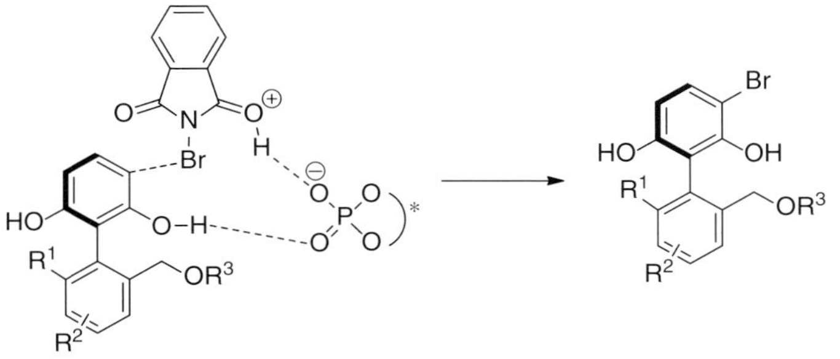

Scheme 7.28 Mechanism of kinetic resolution [40].

Electrophilic bromine sources are perhaps more widely known for their activation of olefins toward facilitating bromocyclizations. Shi showed in 2011 that hydroxy- and amino-alkenes **17a** could undergo enantioselective bromocyclizations in the presence of **(*S*)-PA 16** and *N*-bromosuccinimide (NBS) to yield tetrahydrofurans and tetrahydropyrroles **17b** with modest levels of enantioselectivity (Scheme 7.29) [42].

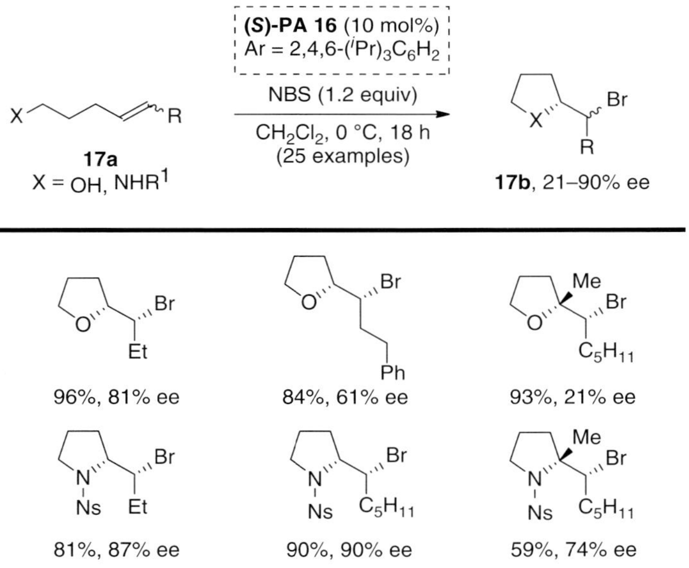

Scheme 7.29 Bromocyclization using alkenes by Shi [42].

The mechanism of the reaction is given in Scheme 7.30. The reaction is proposed to involve a bifunctional catalyst, which hydrogen-bonds to the succinimide carbonyl of NBS and the nucleophile. This allows for selective formation of a bromonium ion, which can be trapped by the pendant nucleophile. This concept has been extended to include sulfur and selenium electrophiles using succinimide-derived reagents and chiral phosphoric acids [43–45]. Closely following Shi's work, Denmark has also reported a bromoetherification with modest results [46].

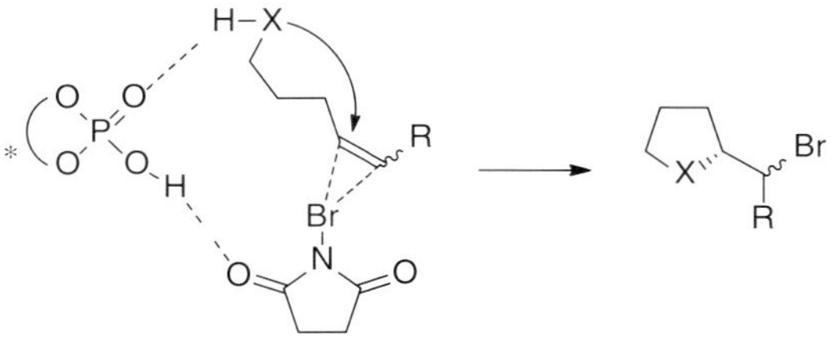

Scheme 7.30 Mechanism of bromocyclization [42].

The intermolecular bromofunctionalization of olefins is a considerably greater challenge than the corresponding intramolecular variant. To date, only one report from Zhang and Tang has been achieved with modest results. They reported on the bromoesterification of cyclic alkenes **18b** using carboxylic acids **18a** in the presence of **PA 4** (Scheme 7.31) [47].

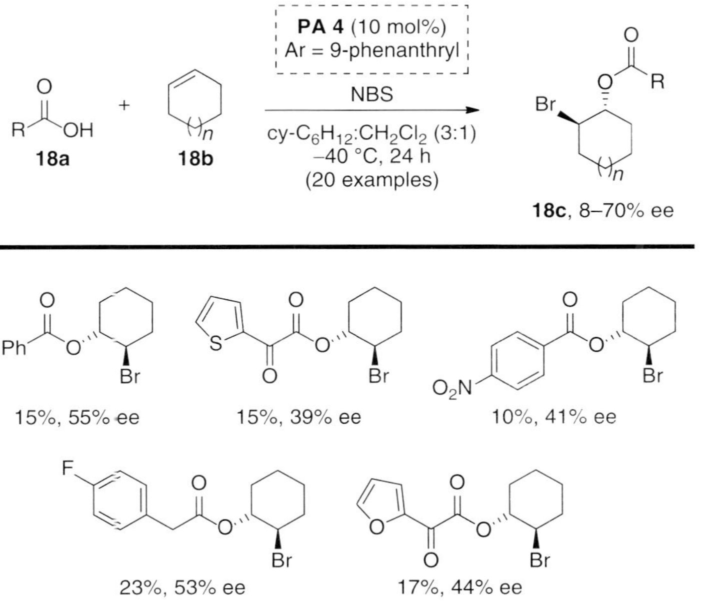

Scheme 7.31 Intermolecular bromoesterification by Zhang and Tang [47].

The reaction proceeds to yield the expected products **18c** with modest enantioselectivities, however, in low yields due to bromoesterification of the catalyst. Nevertheless, the reaction represents an interesting advancement, and further studies in the area should allow for improvements to be made. The mechanism of the reaction can be observed in Scheme 7.32. Reaction of the alkene with NBS generates a bromonium ion with a bound phosphate counterion. While hydrogen-bonding to the carboxylic acid, the catalyst can promote the stereoselective intermolecular attack leading to the products.

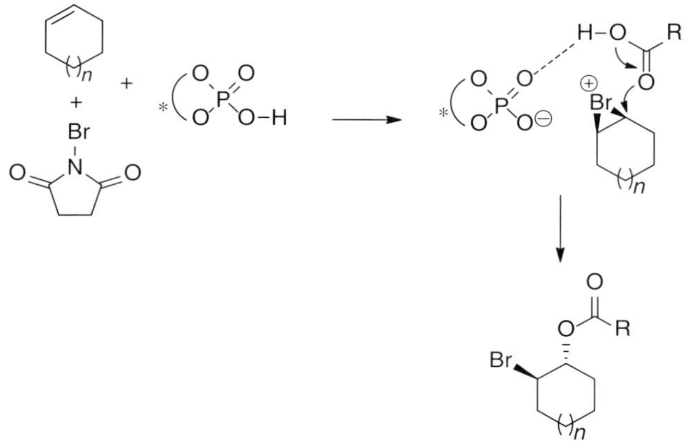

Scheme 7.32 Mechanism of bromoesterification [47].

7.6 Oxidizing Agents

The area of oxidizing agents has not received high levels of attention from the synthetic community, which may be a representation of the challenges that are faced upon attempting to combine these agents with chiral Brønsted acids. One of the early reports in this field was reported in 2005 by Yamamoto, who reported on a nitroso-aldol reaction to access α-oxygenated ketones. He showed that by combining enamines **19a** with nitrosobenzene **19b** in the presence of glycolic acid **BA 23**, a smooth oxygenation took place to yield the products **19c** with high enantioselectivities (Scheme 7.33) [48].

In 2010, Zhong reported a similar procedure using a chiral phosphoric acid catalyst [49], and a year earlier, he developed the process with β-dicarbonyls. He showed that by taking substrates **19d** with 1-chloro-4-nitrosobenzene **19e** and **PA 16**, the desired oxygenation reaction occurred to deliver the products **19f** with generally excellent enantioselectivities (Scheme 7.34) [50]. The mechanism

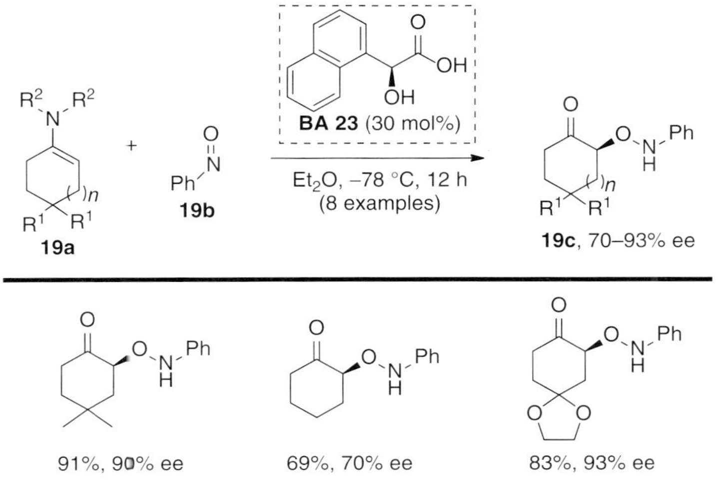

Scheme 7.33 α-oxygenation by Yamamoto [48].

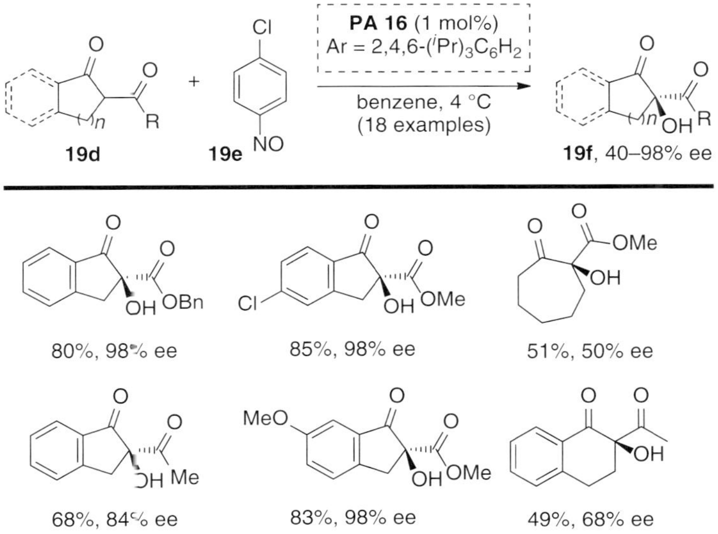

Scheme 7.34 α-oxygenation by Zhong [50].

of Yamamoto's work is shown in Scheme 7.35. The high regioselectivity observed for O- versus N-addition is thought to be due to the catalyst preferring to protonate the nitrogen atom. This activates the nitrosobenzene for attack on the oxygen atom and thus yielding the products after hydrolysis.

Hydrogen peroxide is a highly versatile oxidant, which has been employed for a plethora of oxidative transformations. Since it is usually used as an aqueous solution, the challenge to utilize Brønsted acid catalysts for performing asymmetric transformations is even greater. To date, only a single transformation has

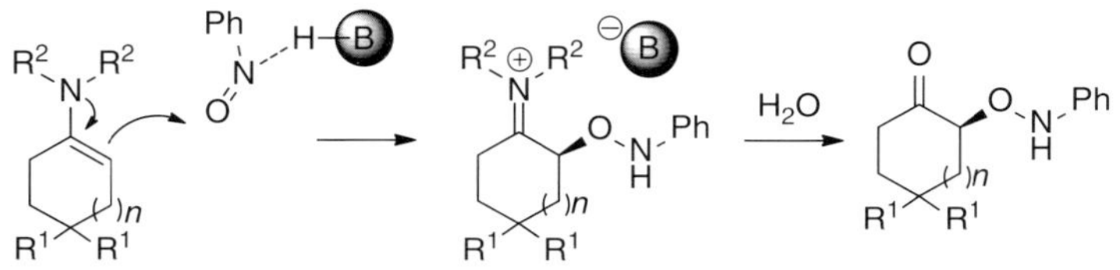

Scheme 7.35 Mechanism of α-oxygenation [48].

been studied using H_2O_2, namely the asymmetric oxidation of sulfides to sulfoxides. The first report came from Tao and Wang in 2012, who employed a chiral phosphoric acid catalyst and managed to achieve modest results [51]. Later that year, List disclosed an improved procedure using an imidodiphosphoric acid catalyst **(*S*)-NPA 11**. This highly confined catalyst was shown to oxidize a variety of alkyl and aromatic sulfides **20a** to the corresponding sulfoxides **20b** with excellent enantioselectivites (Scheme 7.36) [52].

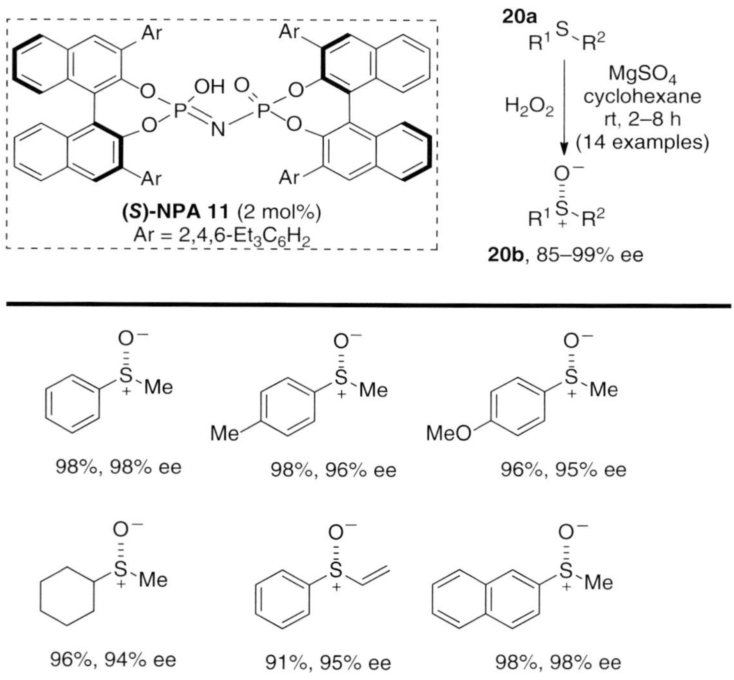

Scheme 7.36 Sulfoxidation by List [52].

The H_2O contained within the oxidant and what was produced during the reaction adversely affected the reaction's efficiency. To alleviate this issue, the group found that the use of $MgSO_4$ as a drying agent severely improved the rate of the reaction and allowed the catalyst loadings to be lowered. An in-depth mechanistic understanding of the catalysts role has also been provided by Sunoj [53].

7.7 Miscellaneous Substrates

The substrates included in this final section are currently employed in single-use cases; however, the principles behind their activation and sense of undergoing enantioselective transformations may be relevant for further developments in this and other fields. The first example is the activation of ketene dithioacetals shown by List in 2012 to be suitable substrates for catalytic asymmetric protonations. He showed that by taking substrates **21a** with either **PA 18** or **PA 40**, a protonative cyclization took place to yield the orthoester **21b** with high enantioselectivities (Scheme 7.37) [54].

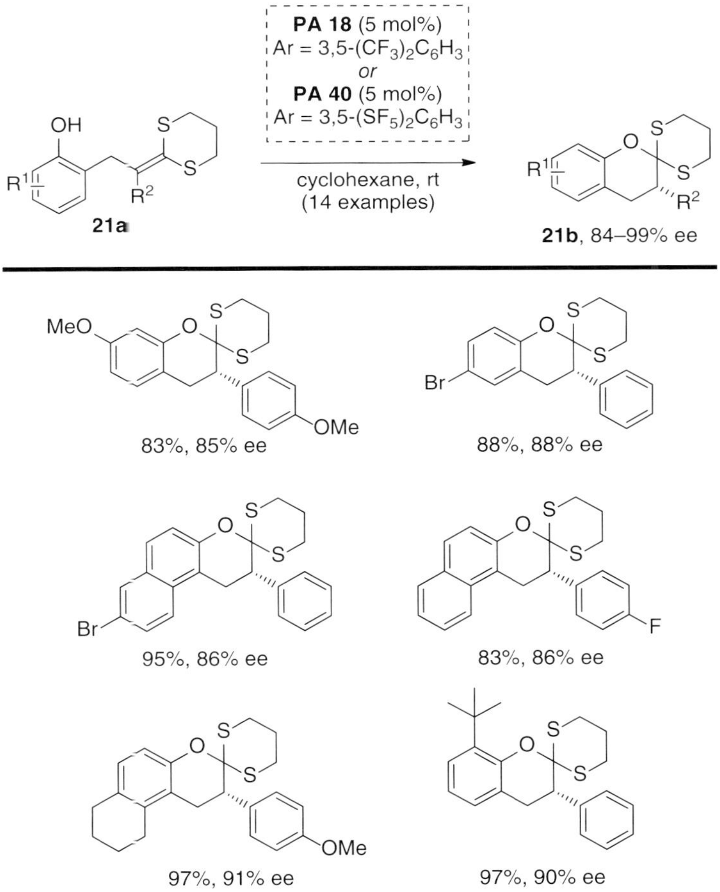

Scheme 7.37 Protonation of ketene dithioacetals by List [54].

The reaction proceeds well for a variety of aromatic groups, and the starting materials are prepared from the corresponding hydrocoumarin precursors. Upon subsequent reaction, the dithioacetal group could be easily hydrolyzed to yield the enantioenriched hydrocoumarin product and thus the transformation can be thought of as a deracemization process. An asymmetric protonation was also shown by the List group a year later for silyl ketene imines. Treatment of substrates **22a** with **PA 16** or **SPA 1** yielded α-alkyl-α-aryl nitriles **22b** with high enantioselectivites (Scheme 7.38) [55].

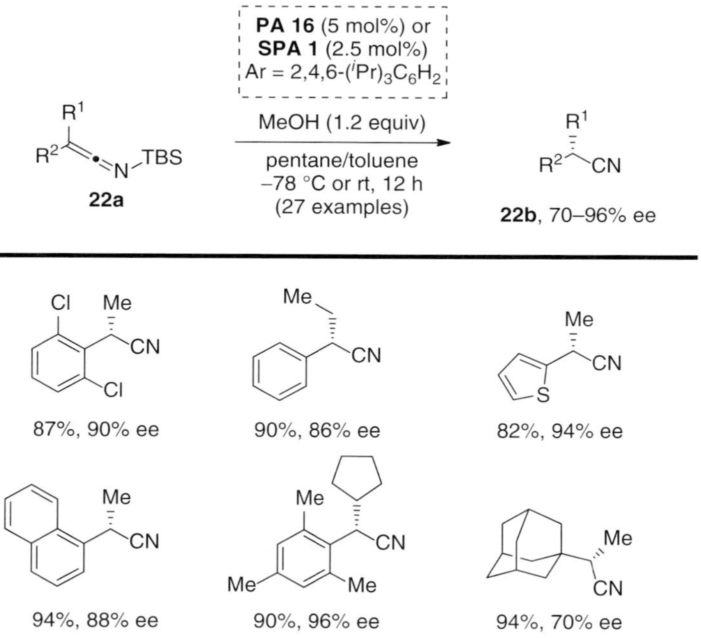

Scheme 7.38 Protonation of silyl ketene imines by List [55].

The stoichiometric proton source is delivered from the addition of methanol, which is also thought to serve as a capture agent for the silyl group upon protonation. The reaction performed equally as well on a gram scale and was demonstrated to be useful to access a structural similar family member to ibuprofen. The mechanism of the process is illustrated in Scheme 7.39. The phosphoric acid catalyst first activates MeOH, which can then coordinate with the silyl group of the silyl ketene imine. This allows for the concerted removal of the silicon group and enantioselective protonation to the yield the desired products.

In the final example in this section, Yamamoto disclosed the activation of nitrones to undergo 1,3-dipolar cycloadditions using a chiral *N*-triflyl phosphoramide catalyst. He showed that aryl nitrones **23a** and enol ether **23b** would undergo a cycloaddition in the presence of **NPA 13** to yield the corresponding isoxazolidines **23c** with good enantioselectivities (Scheme 7.40) [56].

The group found that with a Lewis acid, *exo*-selectivity was observed; however, the Brønsted acid catalyzed pathway favored *endo*-selectivity. Interestingly, the

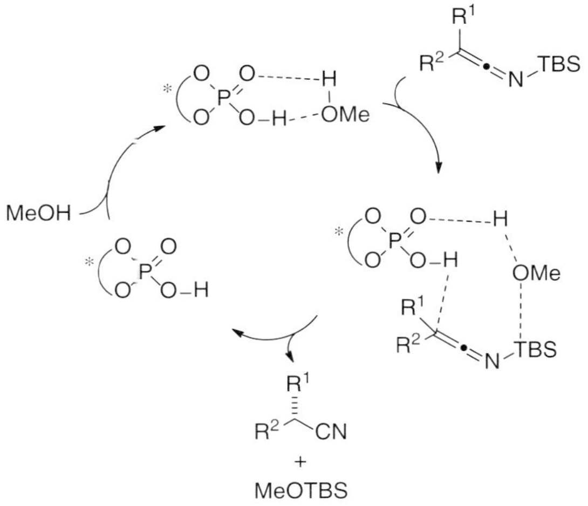

Scheme 7.39 Mechanism of protonation of silyl ketene imines [55].

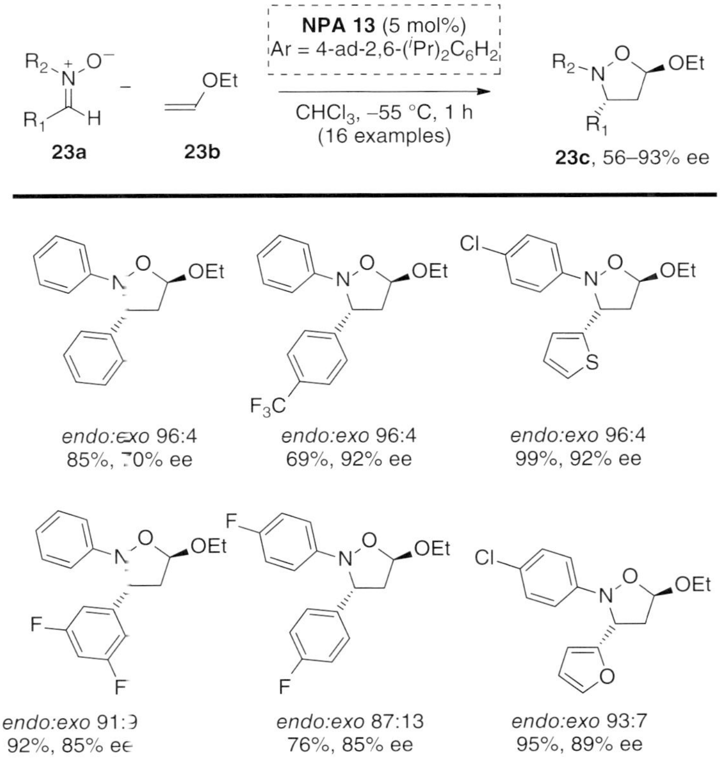

Scheme 7.40 1,3-Dipolar cycloaddition of nitrones by Yamamoto [56].

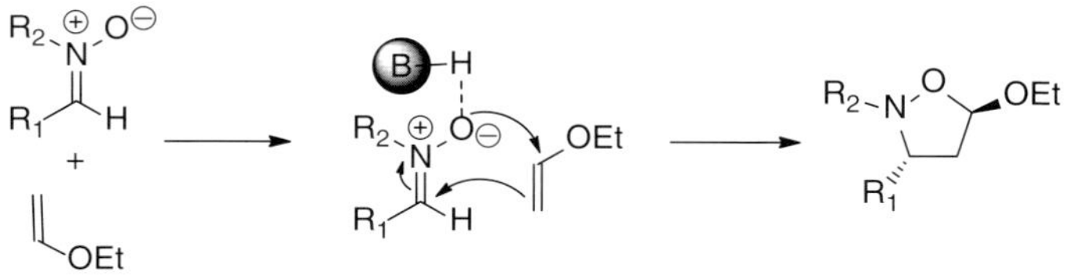

Scheme 7.41 Mechanism of 1,3-dipolar cycloaddition [56].

group found during their catalyst optimizations that especially bulky groups in the 4-position of the aromatic ring greatly improved the enantioselectivity and this may be of consideration for future methodologies. The mechanism is depicted in Scheme 7.41. The catalyst is thought to activate the nitrone by protonation of the basic oxygen atom. While activated, the nitrone can then undergo a stereoselective cycloaddition with the enol ether to yield the desired products and regenerate the catalyst once again. The Brønsted acid is thought to allow for an *endo*-selective reaction, while Lewis acids are known to prefer *exo*-selectivity.

References

1. Rowland, E.B., Rowland, G.B., Rivera-Otero, E., and Antilla, J.C. (2007) *J. Am. Chem. Soc.*, **129**, 12084–12085.
2. Della Sala, G. and Lattanzi, A. (2009) *Org. Lett.*, **11**, 3330–3333.
3. Larson, S.E., Baso, J.C., Li, G., and Antilla, J.C. (2009) *Org. Lett.*, **11**, 5186–5189.
4. Senatore, M., Lattanzi, A., Santoro, S., Santi, C., and Della Sala, G. (2011) *Org. Biomol. Chem.*, **9**, 6205–6207.
5. Monaco, M.R., Poladura, B., Diaz de Los Bernardos, M., Leutzsch, M., Goddard, R., and List, B. (2014) *Angew. Chem. Int. Ed.*, **53**, 7063–7067.
6. Wang, Z., Law, W.K., and Sun, J. (2013) *Org. Lett.*, **15**, 5964–5966.
7. Monaco, M.R., Prévost, S., and List, B. (2014) *J. Am. Chem. Soc.*, **136**, 16982–16985.
8. Monaco, M.R., Prévost, S., and List, B. (2014) *Angew. Chem. Int. Ed.*, **53**, 8142–8145.
9. Wang, Z., Chen, Z., and Sun, J. (2013) *Angew. Chem. Int. Ed.*, **52**, 6685–6688.
10. Chen, Z., Wang, Z., and Sun, J. (2013) *Chem. Eur. J.*, **19**, 8426–8430.
11. Zhuang, M. and Du, H. (2013) *Org. Biomol. Chem.*, **11**, 1460–1462.
12. Čorić, I., Kim, J.H., Vlaar, T., Patil, M., Thiel, W., and List, B. (2013) *Angew. Chem. Int. Ed.*, **52**, 3490–3493.
13. Müller, S., Webber, M.J., and List, B. (2011) *J. Am. Chem. Soc.*, **133**, 18534–18537.
14. Martínez, A., Webber, M.J., Müller, S., and List, B. (2013) *Angew. Chem. Int. Ed.*, **52**, 9486–9490.
15. Kötzner, L., Webber, M.J., Martínez, A., De Fusco, C., and List, B. (2014) *Angew. Chem. Int. Ed.*, **53**, 5202–5205.
16. Li, G.-Q., Gao, H., Keene, C., Devonas, M., Ess, D.H., and Kürti, L. (2013) *J. Am. Chem. Soc.*, **135**, 7414–7417.
17. De, C.K., Pesciaioli, F., and List, B. (2013) *Angew. Chem. Int. Ed.*, **52**, 9293–9295.
18. Zhang, Z. and Antilla, J.C. (2012) *Angew. Chem. Int. Ed.*, **51**, 11778–11782.
19. Nelson, H.M., Reisberg, S.H., Shunatona, H.P., Patel, J.S., and Toste, F.D. (2014) *Angew. Chem. Int. Ed.*, **53**, 5600–5603.
20. Nelson, H.M., Patel, J.S., Shunatona, H.P., and Toste, F.D. (2015) *Chem. Sci.*, **6**, 170–173.
21. Wang, S.-G., Yin, Q., Zhuo, C.-X., and You, S.-L. (2015) *Angew. Chem. Int. Ed.*, **54**, 647–650.

22. Dumoulin, A., Lalli, C., Retailleau, P., and Masson, G. (2015) *Chem. Commun.*, **51**, 5383–5386.
23. Hashimoto, T., Isobe, S., Callens, C.K.A., and Maruoka, K. (2012) *Tetrahedron*, **68**, 7630–7635.
24. So, S.S. and Mattson, A.E. (2014) *Asian J. Org. Chem.*, **3**, 425–428.
25. Rauniyar, V., Lackner, A.D., Hamilton, G.L., and Toste, F.D. (2011) *Science*, **334**, 1681–1684.
26. Phipps, R.J. and Toste, F.D. (2013) *J. Am. Chem. Soc.*, **135**, 1268–1271.
27. Wang, Y.-M., Wu, J., Hoong, C., Rauniyar, V., and Toste, F.D. (2012) *J. Am. Chem. Soc.*, **134**, 12928–12931.
28. Xie, W., Jiang, G., Liu, H., Hu, J., Pan, X., Zhang, H., Wan, X., Lai, Y., and Ma, D. (2013) *Angew. Chem. Int. Ed.*, **52**, 12924–12927.
29. Liu, H., Jiang, G., Pan, X., Wan, X., Lai, Y., Ma, D., and Xie, W. (2014) *Org. Lett.*, **16**, 1908–1911.
30. Phipps, R.J., Hiramatsu, K., and Toste, F.D. (2012) *J. Am. Chem. Soc.*, **134**, 8376–8379.
31. Honjo, T., Phipps, R.J., Rauniyar, V., and Toste, F.D. (2012) *Angew. Chem. Int. Ed.*, **51**, 9684–9688.
32. Shunatona, H.P., Früh, N., Wang, Y.-M., Rauniyar, V., and Toste, F.D. (2013) *Angew. Chem. Int. Ed.*, **52**, 7724–7727.
33. Wu, J., Wang, Y.-M., Drljevic, A., Rauniyar, V., Phipps, R.J., and Toste, F.D. (2013) *Proc. Natl. Acad. Sci. U.S.A.*, **110**, 13729–13733.
34. Zi, W., Wang, Y.-M., and Toste, F.D. (2014) *J. Am. Chem. Soc.*, **136**, 12864–12867.
35. Romanov-Michailidis, F., Guénée, L., and Alexakis, A. (2013) *Angew. Chem. Int. Ed.*, **52**, 9266–9270.
36. Romanov-Michailidis, F., Pupier, M., Guénée, L., and Alexakis, A. (2014) *Chem. Commun.*, **50**, 13461–13464.
37. Romanov-Michailidis, F., Pupier, M., Besnard, C., Bürgi, T., and Alexakis, A. (2014) *Org. Lett.*, **16**, 4988–4991.
38. Romanov-Michailidis, F., Guénée, L., and Alexakis, A. (2013) *Org. Lett.*, **15**, 5890–5893.
39. Mori, K., Ichikawa, Y., Kobayashi, M., Shibata, Y., Yamanaka, M., and Akiyama, T. (2013) *J. Am. Chem. Soc.*, **135**, 3964–3970.
40. Mori, K., Ichikawa, Y., Kobayashi, M., Shibata, Y., Yamanaka, M., and Akiyama, T. (2013) *Chem. Sci.*, **4**, 4235–4239.
41. Mori, K., Kobayashi, M., Itakura, T., and Akiyama, T. (2015) *Adv. Synth. Catal.*, **357**, 35–40.
42. Huang, D., Wang, H., Xue, F., Guan, H., Li, L., Peng, X., and Shi, Y. (2011) *Org. Lett.*, **13**, 6350–6353.
43. Li, L., Li, Z., Huang, D., Wang, H., and Shi, Y. (2013) *RSC Adv.*, **3**, 4523–4525.
44. Guan, H., Wang, H., Huang, D., and Shi, Y. (2012) *Tetrahedron*, **68**, 2728–2735.
45. Wei, Q., Wang, Y.-Y., Du, Y.-L., and Gong, L.-Z. (2013) *Beilstein J. Org. Chem.*, **9**, 1559–1564.
46. Denmark, S.E. and Burk, M.T. (2011) *Org. Lett.*, **14**, 256–259.
47. Li, G.-X., Fu, Q.-Q., Zhang, X.-M., Jiang, J., and Tang, Z. (2012) *Tetrahedron: Asymmetry*, **23**, 245–251.
48. Momiyama, N. and Yamamoto, H. (2005) *J. Am. Chem. Soc.*, **127**, 1080–1081.
49. Lu, M., Lu, Y., Zhu, D., Zeng, X., Li, X., and Zhong, G. (2010) *Angew. Chem. Int. Ed.*, **49**, 8588–8592.
50. Lu, M., Zhu, D., Lu, Y., Zeng, X., Tan, B., Xu, Z., and Zhong, G. (2009) *J. Am. Chem. Soc.*, **131**, 4562–4563.
51. Liu, Z.-M., Zhao, H., Li, M.-Q., Lan, Y.-B., Yao, Q.-B., Tao, J.-C., and Wang, X.-W. (2012) *Adv. Synth. Catal.*, **354**, 1012–1022.
52. Liao, S., Čorić, I., Wang, Q., and List, B. (2012) *J. Am. Chem. Soc.*, **134**, 10765–10768.
53. Jindal, G. and Sunoj, R.B. (2014) *Angew. Chem. Int. Ed.*, **53**, 4432–4436.
54. Lee, J.-W. and List, B. (2012) *J. Am. Chem. Soc.*, **134**, 18245–18248.
55. Guin, J., Varseev, G., and List, B. (2013) *J. Am. Chem. Soc.*, **135**, 2100–2103.
56. Jiao, P., Nakashima, D., and Yamamoto, H. (2008) *Angew. Chem. Int. Ed.*, **47**, 2411–2413.

Experimental Protocols

Appendix A: Catalyst Frequency

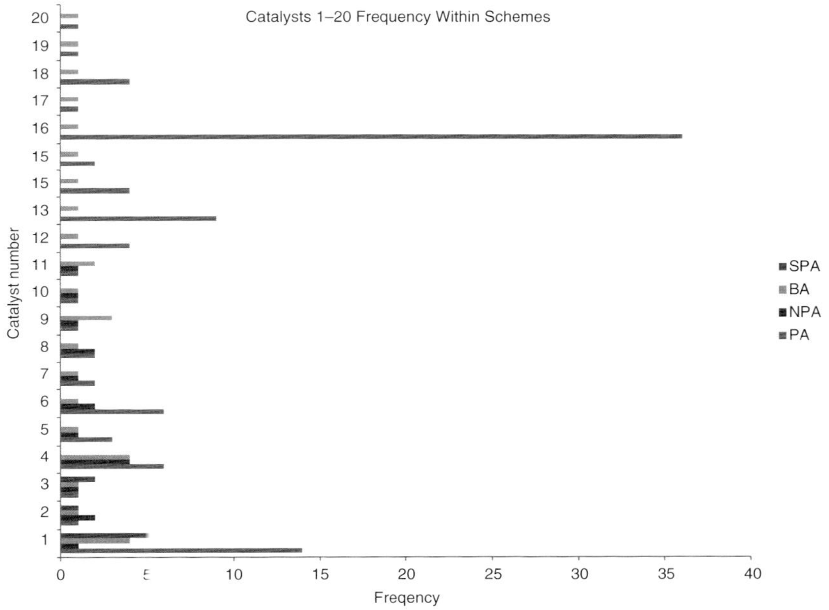

Asymmetric Brønsted Acid Catalysis, First Edition. Magnus Rueping, Dixit Parmar and Erli Sugiono.

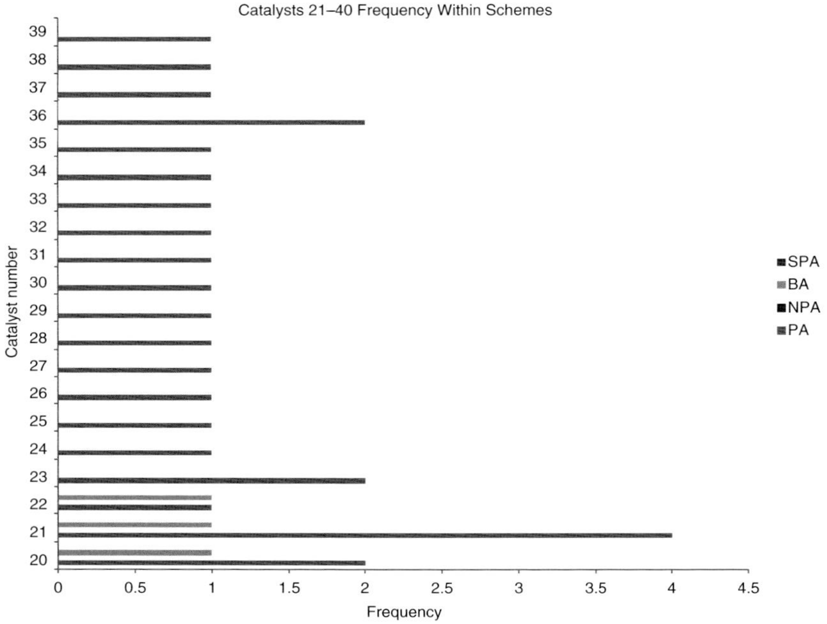
Catalysts 21–40 Frequency Within Schemes
Catalyst number
39
38
37
36
35
34
33
32
31
30
29
28
27
26
25
24
23
22
21
20
0
0.5
1
1.5
2
2.5
3
3.5
4
4.5
Frequency
SPA
BA
NPA
PA

Appendix B: Overview of Phosphoric Acids (PA)

Table B.1 List of **PA** catalysts and example usage.

Catalyst		Example usage
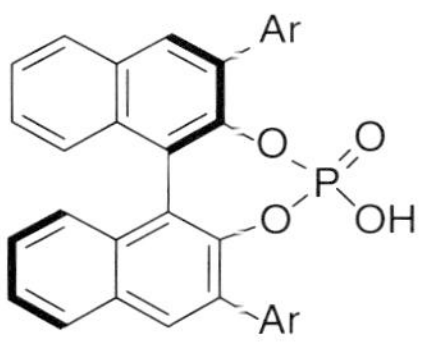	Ar = 9-Anthryl (**PA 1**) Ar = 9-Phenanthryl (**PA 4**) Ar = 1-Naphthyl (**PA 14**) Ar = 2-Naphthyl (**PA 22**) Ar = 9-Pyren-1-yl (**PA 33**)	**PA 1:** additions to imines, transfer hydrogenation, aza-Diels–Alder, electrocyclizations, reduction of ketones, Friedel–Crafts **PA 4:** hydrazone addition to imines, Strecker, transfer hydrogenation **PA 14:** Friedel–Crafts, additions to 3-hydroxy indoles, pinacol rearrangement **PA 22:** kinetic resolution, aza-Diels–Alder, aza-Cope **PA 33:** Baeyer–Villiger
Ar = C_6F_5 (**PA 3**) Ar = 4-$NO_2C_6H_4$ (**PA 8**) Ar = 2-(Np)C_6H_4 (**PA 9**) Ar = C_6H_5 (**PA 12**) Ar = 4-PhC_6H_4 (**PA 15**) Ar = 4-ClC_6H_4 (**PA 24**) Ar = 4-OMeC_6H_4 (**PA 25**)	Ar = 3,5-(Mesityl)$_2C_6H_3$ (**PA 17**) Ar = 3,5-$(CF_3)_2C_6H_3$ (**PA 18**) Ar = 3,5-$Ph_2C_6H_3$ (**PA 27**) Ar = 3,5-$(SF_5)_2C_6H_3$ (**PA 40**)	**PA 3:** allylation of imines **PA 8:** Mannich, self-coupling enamines **PA 9:** Mannich **PA 12:** Biginelli, additions to 3-hydroxy indoles, Diels–Alder **PA 15:** Friedel–Crafts, additions to 3-vinyl indoles **PA 24:** aza-Diels–Alder **PA 25:** additions to 3-hydroxy indoles
Ar = Si(4-*t*Bu$C_6H_4)_3$ (**PA 2**) Ar = $SiPh_3$ (**PA 13**)	Ar = 2,4,6-(Me)$_3C_6H_2$ (**PA 5**) Ar = 2,4,6-(^{i}Pr)$_3C_6H_2$ (**PA 16**) Ar = 2,4,6-(Cy)$_3C_6H_2$ (**PA 38**) Ar = 2,4,6-(*c*-Pent)$_3C_6H_2$ (**PA 39**)	**PA 2:** aza-Darzens **PA 13:** Biginelli, Friedel–Crafts, Pictet–Spengler, transfer hydrogenation, hydroamination **PA 17:** Friedel–Crafts **PA 18:** transfer hydrogenation **PA 27:** additions to hemiaminals

Asymmetric Brønsted Acid Catalysis, First Edition. Magnus Rueping, Dixit Parmar and Erli Sugiono.

Table B.1 (Continued)

Catalyst	Example usage
	PA 40: enantioselective protonation **PA 5:** isocyanide addition to imines, vinylogous Mannich, aldol **PA 16:** Friedel–Crafts, Pictet–Spengler, transfer hydrogenation, additions to hemiaminals, aza-Petasis–Ferrier, allylations/propargylations, Robinson annulations, aldol, transacetalization, desymmetrizations **PA 38:** phase-transfer fluorination **PA 39:** phase-transfer fluorination
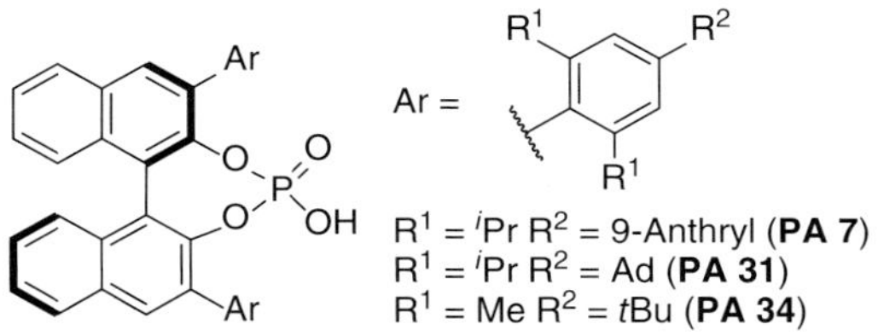	**PA 7:** amine addition to imine, Kabachnik–Fields **PA 31:** Friedel–Crafts of ketones **PA 34:** additions to *o*-quinone methides
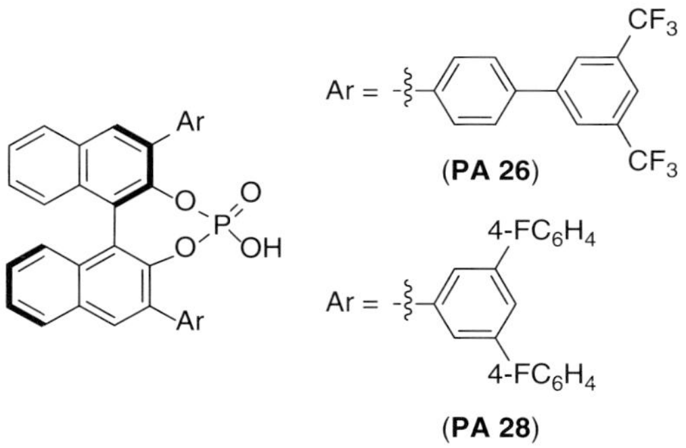	**PA 26:** Friedel–Crafts **PA 28:** additions to aminals
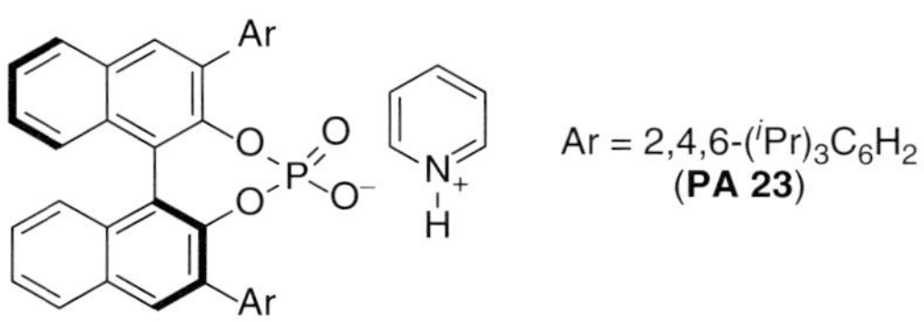	**PA 23:** aza-Diels–Alder

Table B.1 (Continued)

Catalyst	Example usage
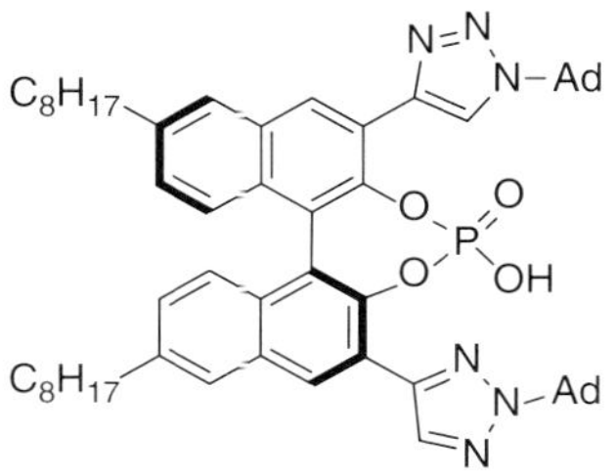 (**PA 30**)	**PA 30:** cross-dehydrogenative couplings
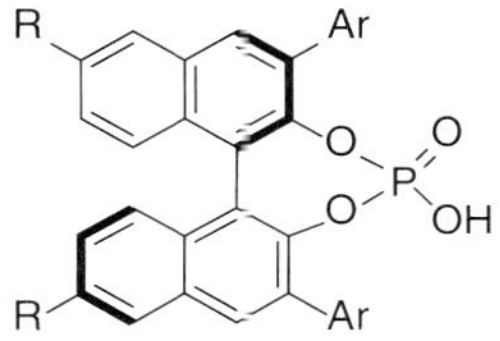 Ar = 2,4,6-(^{i}Pr)$_3$C$_6$H$_2$ R = NO$_2$ (**PA 32**) R = Ph$_3$P$^+$PF$_6^-$ (**PA 35**) R = C$_8$H$_{17}$ (**PA 37**)	**PA 32:** kinetic resolution **PA 35:** Friedel–Crafts of alkenes **PA 37:** phase-transfer fluorination
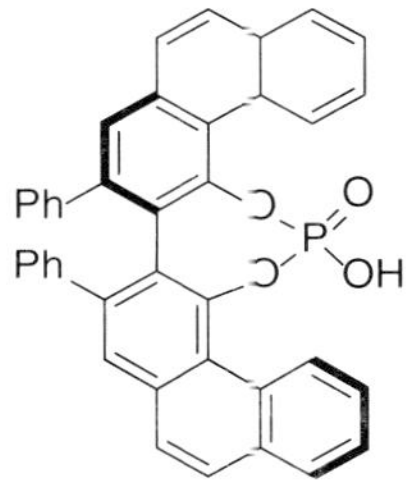 (**PA 6**)	**PA 6:** amine addition to imines, transfer hydrogenation, aziridine openings
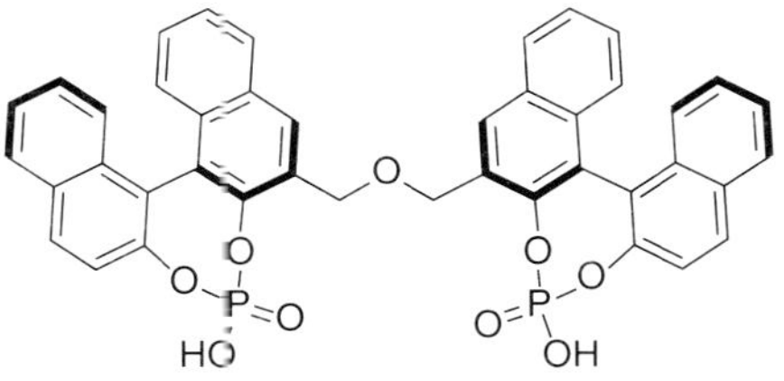 (**PA 21**)	**PA 21:** cascade hydride transfer, 1,3-dipolar cycloaddition

Table B.1 (*Continued*)

Catalyst	Example usage
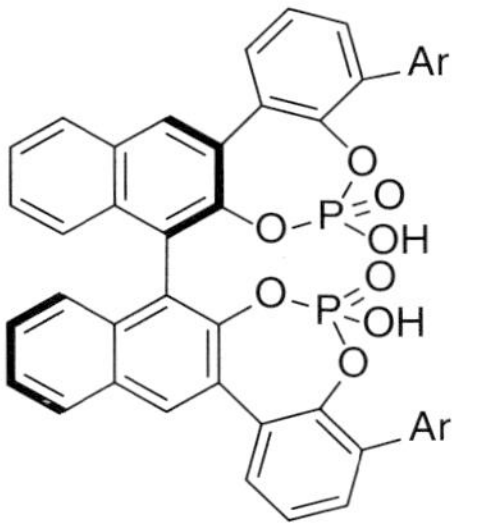 Ar = 2,4,6-(^{i}Pr)$_3C_6H_2$ (**PA 36**)	**PA 36:** Diels–Alder
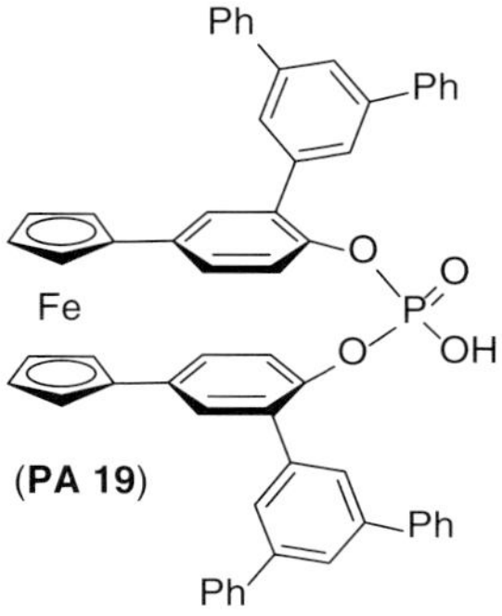 (**PA 19**) 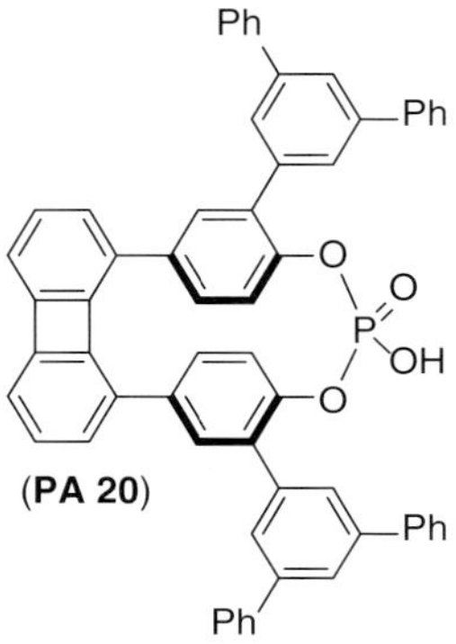 (**PA 20**)	**PA 19/20:** transfer hydrogenation
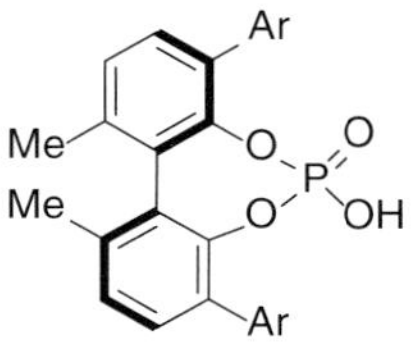 Ar = 4-$NO_2C_6H_4$ (**PA 10**) Ar = 9-Anthryl (**PA 11**) Ar = 2,4-$(CF_3)_2C_6H_3$ (**PA 29**)	**PA 10/11:** Mannich **PA 29:** hydride shift

Appendix C: Overview of *N*-Phosphoramide Acids (NPA)

Catalyst	Example usage
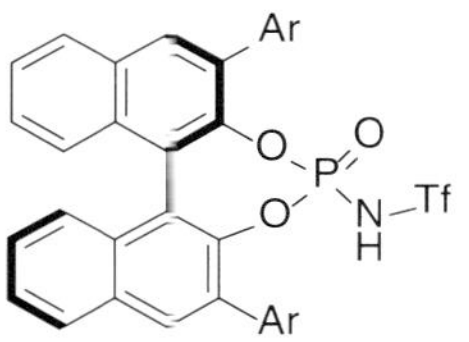 Ar = $SiPh_3$ (**NPA 6**) Ar = 2-Naphthyl (**NPA 7**) Ar = 9-Phenanthryl (**NPA 9**) Ar = 4-$NO_2C_6H_4$ (**NPA 3**) Ar = Ph (**NPA 12**) Ar = 4-$^tBuC_6H_4$ (**NPA 10**) Ar = 2,4,6-$(^iPr)_3C_6H_2$ (**NPA 4**) Ar = 4-ad-2,6-$(^iPr)_2C_6H_2$ (**NPA 13**)	**NPA 6:** indole additions to hydroxy-lactams, Friedel–Crafts **NPA 7:** carbonyl-ene reaction **NPA 9:** Nazarov **NPA 3:** Friedel–Crafts **NPA 12:** intramolecular addition to carbocations **NPA 10:** Nazarov **NPA 4:** Friedel–Crafts, reduction of imines, [3+2]-cycloadditions, radical additions, Schmidt, Diels–Alder **NPA 13:** 1,3-dipolar cycloadditions
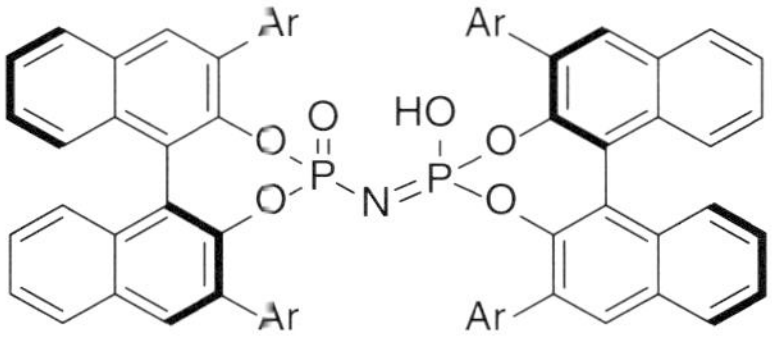 Ar = 1-Naphthyl (**NPA 2**) Ar = 2-^iPr-5-Me-C_6H_3 (**NPA 8**) Ar = 2,4,6-$Et_3C_6H_2$ (**NPA 11**)	**NPA 2:** Mannich, Friedel–Crafts **NPA 8:** desymmetrization of bis-lactones, acetalization of aldehydes **NPA 11:** spiroacetalization, sulfide oxidation

Asymmetric Brønsted Acid Catalysis, First Edition. Magnus Rueping, Dixit Parmar and Erli Sugiono.

Catalyst	Example usage
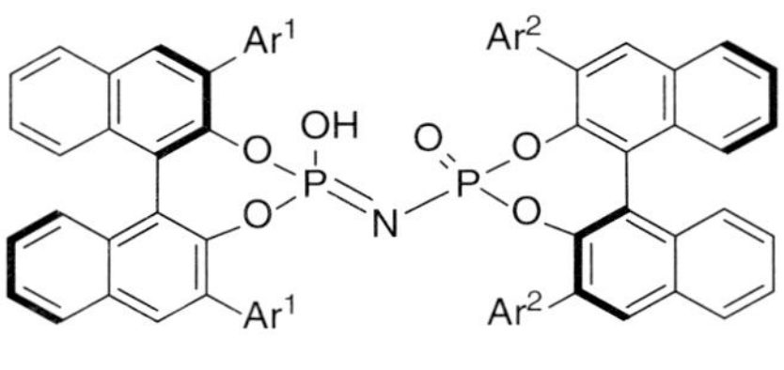 (**NPA 5**) Ar^1 = Phenyl, Ar^2 = 1-Naphthyl	**NPA 5:** pyrrole additions to 3-hydroxy indoles
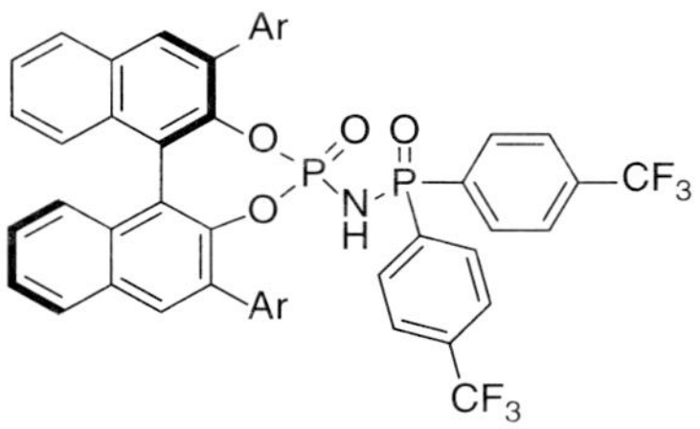 (**NPA 1**) Ar = 2,4,6-$(^i Pr)_3C_6H_2$	**NPA 1:** intramolecular addition of phenols to imines

Appendix C: Overview of SPINOL Phosphoric Acids (SPA)

Catalyst	Example usage
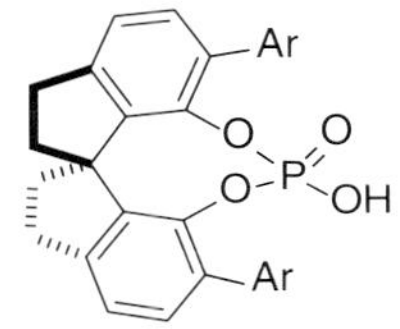 Ar = 2,4,6-(^{i}Pr)$_2$C$_6$H$_2$ (**SPA 1**) Ar = 3,5-(CF$_3$)C$_6$H$_3$ (**SPA 2**) Ar = 9-Anthryl (**SPA 3**)	**SPA 1:** multicomponent aza-Diels–Alder, allylation, protonation of silyl ketene imines, hydride shift (***N*-SPA 1**) **SPA 2:** multicomponent aza-Diels–Alder **SPA 3:** desymmetrization of oxetanes, Fischer-indole synthesis

Asymmetric Brønsted Acid Catalysis, First Edition. Magnus Rueping, Dixit Parmar and Erli Sugiono.

Appendix E Overview of All Other Brønsted Acids (BA)

Catalyst		Example usage
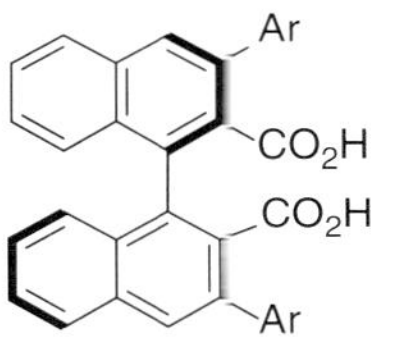	Ar = 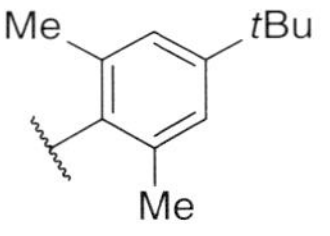 (**BA 1**)	**BA 1:** addition to imines, Strecker, addition to quinone imine ketals
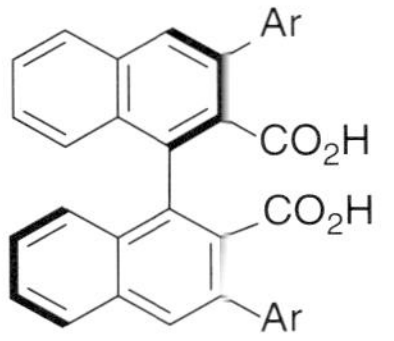	Ar = 2,4,6-$(Me)_3C_6H_2$ (**BA 2**) Ar = $CH(2\text{-}Np)_2$ (**BA 11**) Ar = 3,5-$(NO)_2$-C_6H_3 (**BA 22**)	**BA 2:** aza-Darzens **BA 11:** 1,3-dipolar cycloaddition **BA 22:** semi-pinacol
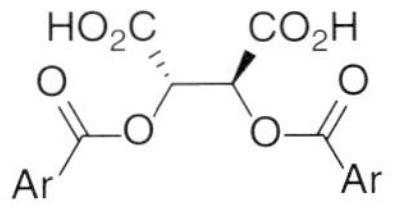	Ar = 4-MeC_6H_4 (**BA 10**)	**BA 10:** Friedel–Crafts
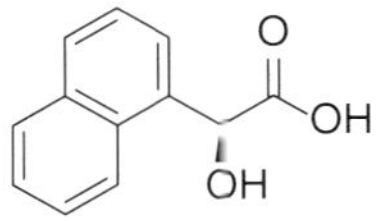	**BA 23** (30 mol%)	**BA 23:** oxygenation
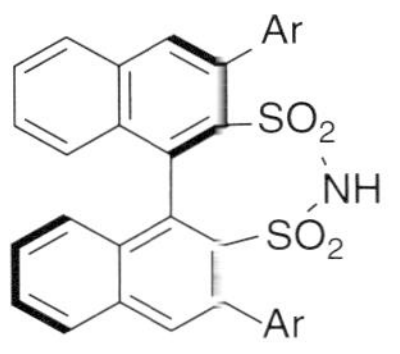	Ar = 4-$CF_3C_6H_4$ (**BA 3**) Ar = 3,5-$(CF_3)_2C_6H_3$ (**BA 9**) Ar = 3,5-$(NO_2)_2$-4-MePh (**BA 12**)	**BA 3:** Hosomi–Sakurai **BA 9:** Friedel–Crafts, Abramov, Aldol **BA 12:** Hosomi–Sakurai

Asymmetric Brønsted Acid Catalysis, First Edition. Magnus Rueping, Dixit Parmar and Erli Sugiono.

Catalyst		Example usage
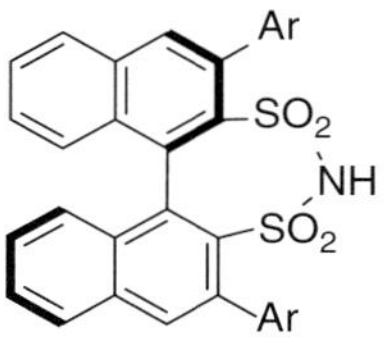	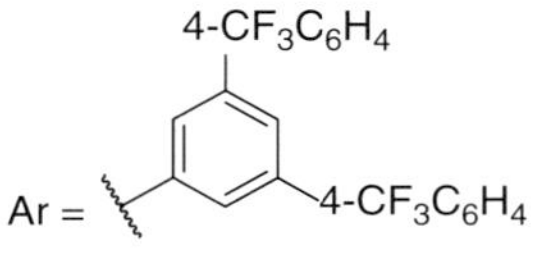 (**BA 6**)	**BA 6:** Mannich
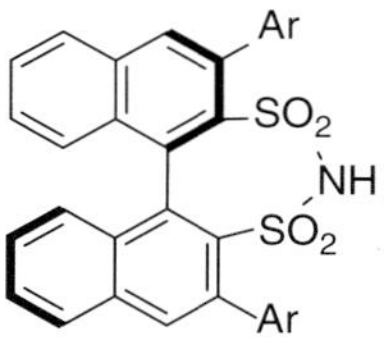	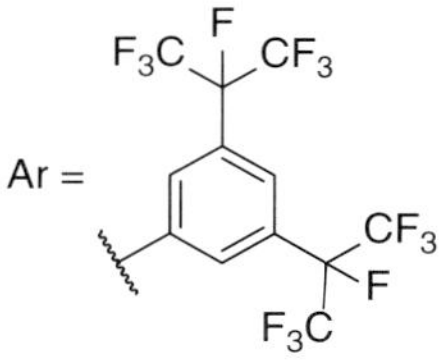 (**BA 15**)	**BA 15:** Diels–Alder
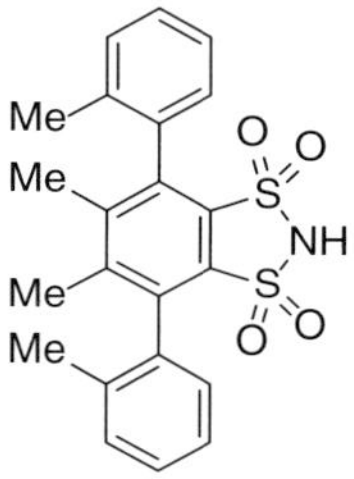	(**BA 8**)	**BA 8:** Strecker
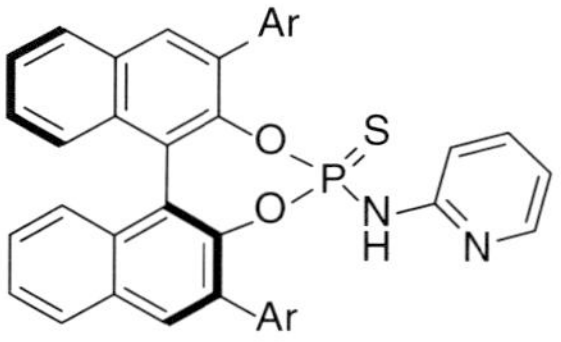	Ar = 3,5-$(CF_3)_2C_6H_3$ (**BA 13**)	**BA 13:** desymmetrization
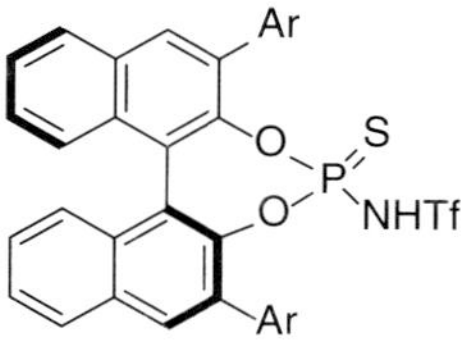	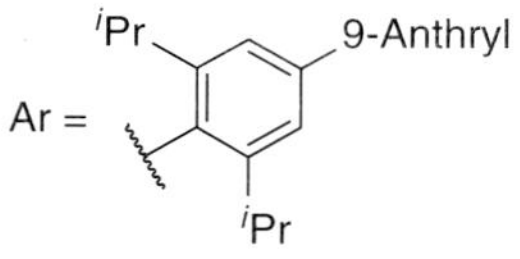 (**BA 14**)	**BA 14:** Aldol
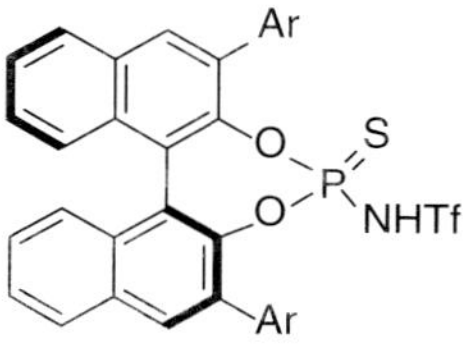	Ar = 4-*t*Bu-2,6-$(^i\mathrm{Pr})_2C_6H_2$ (**BA 16**)	**BA 16:** enantioselective protonation

Catalyst		Example usage
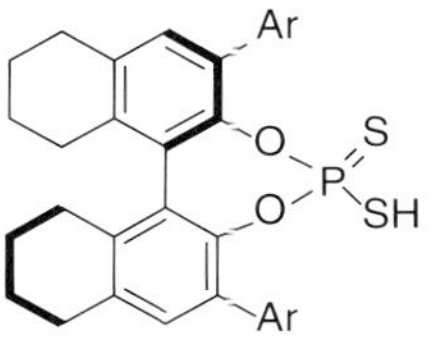	Ar = 10-(3,5-(tBu)$_2$C$_6$H$_3$)-9-anthryl (**BA 18**)	**BA 18:** hydroamination
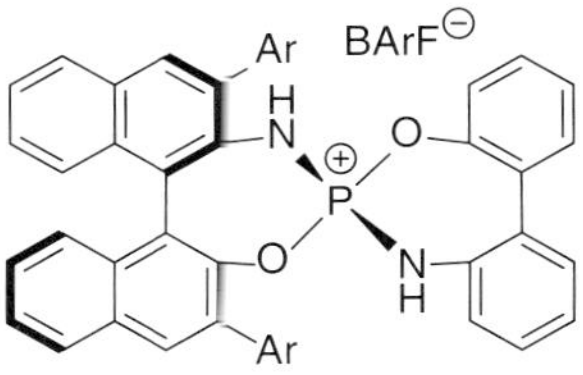	Ar = Ph (**BA 17**)	**BA 17:** enantioselective protonation
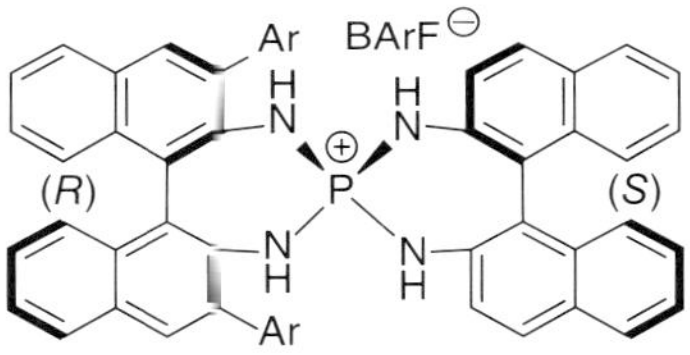	Ar = 3,4,5-F$_3$C$_6$H$_2$ (**BA 19**)	**BA 19:** hydroamination
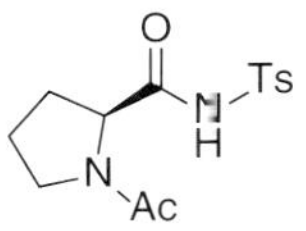	(**BA 4**)	**BA 4:** Kabachnik–Fields
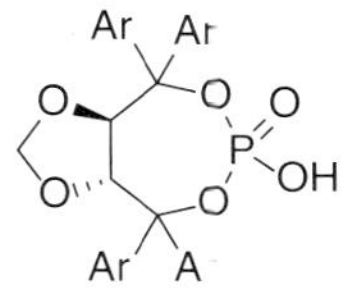	Ar = 4-CF$_3$C$_6$H$_4$ (**BA 5**)	**BA 5:** Mannich
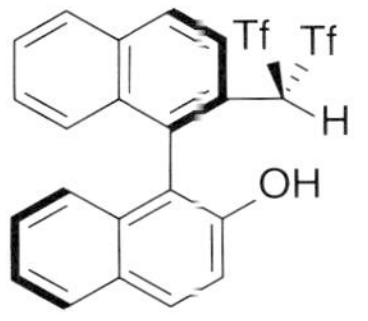	(**BA 7**)	**BA 7:** Mannich

Catalyst		Example usage
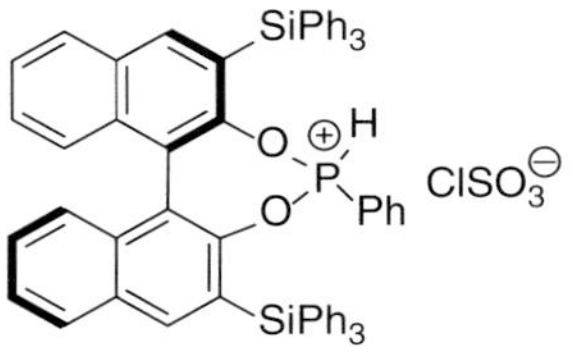	(**BA 20**)	**BA 20:** kinetic resolution **BA 21:** cascade reaction
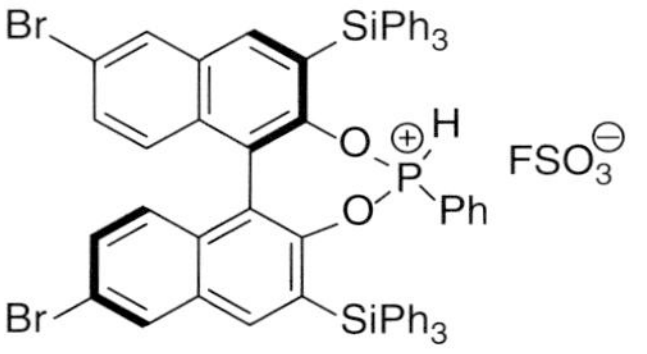	(**BA 21**)	

Index